Digital Signal Processing

Digital Signal Processing
A Practitioner's Approach

Kaluri Venkata Rangarao
International Institute of Information Technology, Hyderabad

Ranjan K Mallik
Indian Institute of Technology, New Delhi

John Wiley & Sons, Ltd

Other Wiley Editorial Offices

John Wiley & Sons Inc., 111 River Street, Hoboken, NJ 07030, USA

Jossey-Bass, 989 Market Street, San Francisco, CA 94103-1741, USA

Wiley-VCH Verlag GmbH, Boschstr. 12, D-69469 Weinheim, Germany

John Wiley & Sons Australia Ltd, 42 McDougall Street, Milton, Queensland 4064, Australia

John Wiley & Sons (Asia) Pte Ltd, 2 Clementi Loop #02-01, Jin Xing Distripark, Singapore 129809

John Wiley & Sons Canada Ltd, 22 Worcester Road, Etobicoke, Ontario, Canada M9W 1L1

Wiley also publishes its books in a variety of electronic formats. Some content that appears in
print may not be available in electronic books.

Library of Congress Cataloging-in-Publication Data

British Library Cataloguing in Publication Data

A catalogue record for this book is available from the British Library

ISBN-13 978-0-470-01769-2 (HB)
ISBN-10 0-470-01769-4 (HB)

Typeset in 10/12pt Times Roman by Thomson Press (India) Limited, New Delhi
Printed and bound in Great Britain by TJ International Ltd, Padstow, Cornwall
This book is printed on acid-free paper responsibly manufactured from sustainable forestry
in which at least two trees are planted for each one used for paper production.

This book is dedicated to
Byrraju Ramalinga Raju, Chairman
Byrraju Rama Raju, Managing Director
K Kalyanrao, President & Chief Technology Officer
Satyam Computer Services Ltd, Hyderabad, India

Contents

Foreword

This foreword is by V. J. Sundaram, former director of DRDL and RCI, Hyderabad; former project director of Prithvi; and former chairman of the programe management board for the Integrated Guided Missile Development Programe (IGMDP), DRDO, India

A Saturn V rocket blasts off and puts a man on the moon. The launch signals are of prime concern to aeronautical, mechanical, electrical and control engineers.

Thirty-five years later the Christmas spirit is shattered by a tsunami. Geologists and civil engineers scramble to study the earthquake signals. Though the events are totally different, the signal structures and analysis methods are similar. Before a surgeon picks up his knife, he asks for an electrocardiogram. A policeman turns to speech identification and biometrics to catch a kidnapper. A nanobot coursing through a human body emits signals to identify itself. Cricket captains scrutinise satellite images for signs of rain. Between the moon mission and the Christmas tsunami, signal processing has made giant strides, with the ubiquitous processor and embedded systems. Dow Jones could never have imagined the invasion of the business world by a tiny DSP with clock speeds going from megahertz to gigahertz.

Dr Rangarao has rich experience in systems – naval, missile, communications, business and education – covering both hardware and software, including full life cycle product development. He has aimed to bring out the basic commonality underlying efficient signal analysis in various fields, even though different terms may sometimes be used for the same operation. He has indicated how design cycle times can be reduced with MATLAB and co-simulation using DSPs, to enable faster product marketing without sacrificing quality and reliability.

The aim of a book, apart from comprehensive coverage, is to ignite the mind with a thirst for more. This Dr Rangarao has achieved with his practitioner's approach.

Lt. Gen. (Retd) Dr V. J. Sundaram PVSM, AVSM, VSM

Hyderabad, April 2005

Preface

There are many books on the subject of digital signal processing, but I have not come across any that cater for practising engineers and beginners. This book aims to meet the needs of people who wish to grasp the concepts and implement them. Without compromising the analytical rigour behind the concepts, it takes you smoothly from grasping the ideas to implementing them for yourself.

Digital signal processing (DSP) is practised by people of all disciplines. Historically, seismologists did major research using DSP, particularly on earthquake prediction. Statisticians such as Box and Jenkins used DSP to study the flow of rivers in North America, finding that they provided an easy way for merchants to transport logs from the mountain forests to the foothills.

Only recently have communications engineers begun using DSP. And it's even more recently that DSP has been embraced by military personnel, space scientists, bankers, stockbrokers, clinical doctors and astronomers, with the famous example of predicting sunspots. In his April 1987 paper 'Challenges to Control in Signal Processing and Communications' (*Control Systems Magazine*, **7**(2), 3–6) Willsky has presented an interesting perspective on unifying the concepts of control, communication, statistics and systems engineering. Enthusiastic readers are advised to consult it.

I have written this book with the mindset of a hardcore electrical engineer, always attempting to translate information into a piece of hardware worthy of implementation. The aim is to make the understanding of DSP as simple as possible as well as readable by beginners and experts alike. I chose as a role model India's great creative genius and scholar Annamayya, who bundled all the complex philosophies into simple and lucid language within the understanding of the layman, the common man. Each chapter is self-contained and can be read on its own.

This book is a compendium of several lectures I have delivered on this subject to a variety of audiences during a decade at the Naval Laboratories in Visakapatnam, two years at the Naval Postgraduate School in California, five years at the Electronics Research Laboratory in Hyderabad, five years at the Research Center

Imarat, five years of industry experience at Satyam Computer Services Ltd, Hyderabad, and one year with General Electric at Begumpet, Hyderabad.

It has been refined over the years by many in-depth discussions with associates, professional colleagues and distinguished persons. I also owe a lot to my master's degree students, who came to me for guidance with their dissertations.

Many lunch and teatime discussions have enriched my perspective on the subject. These experiences have left lasting memories, so much so that I felt it was my duty to write this book, and within my limited wisdom and understanding, to take the things I had learned and pass them on in my own language, blended with my own experience.

This book has six chapters. Chapter 1 introduces the processing of signals. Chapter 2 revisits the basics for understanding DSP. Chapter 3 is on digital filters. Chapter 4 deals with fast Fourier transforms. Chapter 5 gives various hardware schemes for implementation and looks at embedded systems; it gives pointers for very large scale integration (VLSI). The case studies in Chapter 6 give a flavour of some applications for practising engineers.

Kaluri Venkata Rangarao

Acknowledgements

I would like to acknowledge my wife, Kaluri Lavanya, originally Pragada Lavanya, without whose help I could not have written this book. She coped with my idiosyncrasies during this tortuous but pleasant journey. The seed of the book took shape when my daughter Kadambari was doing her BSEE project at home; I would like to thank her. My other daughter, Chinni (Soujanya), bought a Sony Vaio laptop with her own earnings for writing this book need to be acknowledged. And I would like to thank others in the family who supported me at a distance.

Oruganti Gopalakrishna fixed *muhurtham* for the start of writing. I thank my student Anurag for proof-reading and Rear Admiral Mohan, who gave the book a technical review as a finishing touch. The beginnings of the book go back to when Dr U. M. S. Murty of Agilent US asked me to give some lectures on digital signal processing for the people around him; I thank him for that. I acknowledge the encouragement of Professor Iyyanki Muralikrishna, Maganti Bhaskar Murty. And I thank Lieutenant-General V. J. Sundaram, with whom I have been associated for the past 30 years; he sent me to the Indian Institute of Technology, Delhi, so I could write this book. I thank Professor V. U. Reddy for his all-round help.

When I took my first job in the private sector, Mr Grandhi Mallikarjunarao let me buy 4TeX software and allowed me to pursue my work. I greatly acknowledge him for that. The main motivation came from the sales staff – I was trying to sell technology. I used to meet them and try to teach them over lunch or dinner, and sometimes in more formal sessions, as I travelled around. I have to acknowledge them for their drive. The Indian Institute of Technology, Delhi, let me stay on its campus and took me into its fold. The Electrical Engineering Faculty treated me like one of their own, especially the communications group. I give sincere thanks to the Electrical Engineering Faculty. I thank Professor Rajeev Sangal for a good understanding and I acknowledge the faculty of the International Institute of Information Technology (IIIT), Hyderabad, for their subtle help. My students Pallavi, Sirisha and Sai did a wonderful job and I thank Vivian Lee of CMU for a nice tennis photograph.

Many well-wishers in many forms have helped me and encouraged me in this professional and social cause.

Kaluri Venkata Rangarao

1

Processing of Signals

Any sequence or set of numbers, either continuous or discrete, defines a signal in the broad sense. Signals originate from various sources. They occur in data processing or share markets, human heartbeats or telemetry signals, a space shuttle or the golden voice of the Indian playback singer Lata Mangeshkar, the noise of a turbine blade or submarine, a ship or instrumented signal inside a missile.

Processing of signals, whether analogue or digital, is a prerequisite to understanding and analysing them. Conventionally, any signal is associated with time. Typically, a one-dimensional signal has the form $x(t)$ and a two-dimensional signal has the form $f(x, y, t)$. Understanding the origin of signals or their source is of paramount importance. In strict mathematical form, a signal is a mapping function from the real line to the real line, or in the case of discrete signals, it is a mapping from the integer line to the real line;[1] and finally it is a mapping from the integer line to the integer line.

Typically the measured signal $\hat{y}(t)$ is different from the emanated signal $y(t)$. This is due to corruption and can be represented as follows:

$$y(t) = \hat{y}(t) + \gamma(t) \qquad \text{in continuous form,} \qquad (1.1)$$

$$y_k = \hat{y}_k + \gamma_k \qquad \text{in discrete form,} \qquad (1.2)$$

where γ is the *unwanted signal*, commonly referred to as noise and most of the time statistical in nature. This is one of the reasons why processing is performed to obtain $\hat{y}_k$ from y_k.

1.1 Organisation of the Book

Chapter 1 describes how analogue signals are converted into numbers and the associated problems. It gives essential principles of converting the analogue signal

---

[1]Time series.

Digital Signal Processing: A Practitioner's Approach K. V. Rangarao and R. K. Mallik
© 2005 John Wiley & Sons, Ltd

to digital form, independent of technology. Also described are the various domains in which the signals are classified and the associated mathematical transformations. Chapter 2 looks at the basic ideas behind the concepts and provides the necessary background to understand them. Chapter 2 will give confidence to the reader to understand the principles of digital filters. Chapter 3 describes commonly used filters along with practical examples. Chapter 4 focuses on Fourier transform techniques plus computational complexities and variants. It also describes frequency domain least squares in the spectral domain. Chapter 5 looks at methodologies of implementing the filters, various types of converters, limitations of fixed points and the need for pipelining. We conclude by comparing the commercially available processors and by looking at implementations for two practical problems: a DSP processor and a hardware implementation using an FPGA.[2] Chapter 6 describes a system-level application in two domains. The MATLAB programs in the appendices may be modified to suit readers' requirements.

1.2   Classification of Signals

Signals can be studied using spectral characteristics and temporal characteristics. Figure 1.1 shows the same signal as a function of time (*temporal*) or as a function of

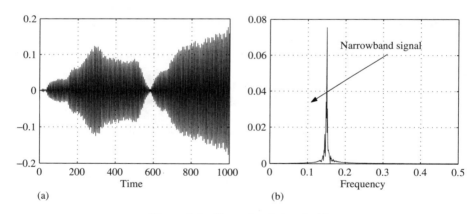

Figure 1.1   Narrowband signal $x(t)$

frequency (*spectral*). It is very important to understand the signals before we process them. For example, the sampling for a narrowband signal will probably be different than the sampling for a broadband signal. There are many ways to classify signals.

---

[2]Field-programable gate array.

1.2.1 Spectral Domain

Signals are generated as a function of time (*temporal* domain) but it is often convenient to analyse them as a function of frequency (*spectral* domain). Figure 1.1 shows a narrowband signal in the temporal domain and the spectral domain. Figure 1.2 does the same for a broadband signal. For historical reasons, signals

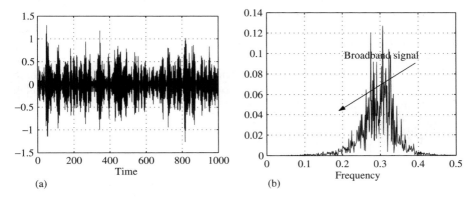

Figure 1.2 Broadband signal $x(t)$

which are classified by their spectral characteristics are more popular. Here are some examples:

1. Band-limited signals
   (a) Narrowband signals
   (b) Wideband signals
2. Additive band-limited signals

1.2.1.1 Band-Limited Signals

Band-limited signals, narrow or wide, are the most commonly encountered signals in real life. They could be a signal from an RF transmitter, the noise of an engine, etc. The signals in Figures 1.1 and 1.2 are essentially the same except that their bandwidths are different and cannot be perceived in the time domain. Only by obtaining the spectral characteristics using the Fourier transform we can distinguish one from the other.

1.2.1.2 Additive Band-Limited Signals

A group of additive band-limited signals are conventionally known as *composite* signals with a wide range of frequencies. Typically, a signal emitted from a radar [1] and its reflection from the target have a carrier frequency of a few gigahertz and a pulse repetition frequency (PRF) measured in kilohertz. The rotation frequency of the radar is a few hertz and the target spatial contour which gets convolved with this signal is measured in fractions of hertz.

   Composite signals are very difficult to process, and demand multi-rate signal processing techniques. A typical composite signal in the time domain is depicted in Figure 1.3(a), which may not give total comprehension. Looking at the spectrum of the same signal in Figure 1.3(b) provides a better understanding. These graphs are merely representative; in reality it is very difficult to sample composite signals due to the wide range of frequencies present at logarithmic distances.

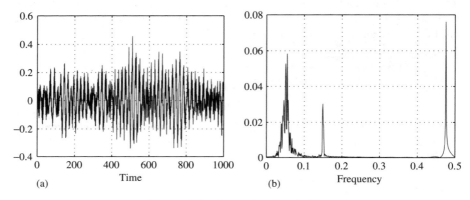

Figure 1.3   Composite signal $x(t)$

1.2.2 Random Signals

Only some signals can be characterised in a deterministic way through their spectral properties. There are some signals which need a probabilistic approach. Random signals such as shown in Figure 1.4(a) can only be characterised by their

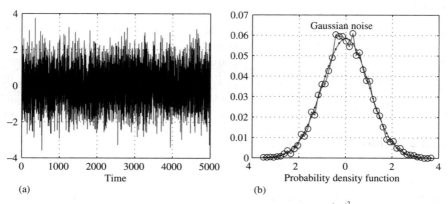

Figure 1.4   Random signal $x(t)$ with pdf $\frac{1}{\sqrt{2\pi}\sigma}e^{\frac{(x-\mu)^2}{2\sigma^2}}$

probability density function (pdf). For the same signal we have estimated the *pdf* by a simple histogram (Figure 1.4(b)). For this signal we have actually generated the pdf from the given *time series* using a numerical technique; notice how closely it matches with the theoretical bell-shaped curve representing a normal distribution.

Statistical or random signals can also be characterised only by their moments, such as first moment (mean μ), second moment (variance σ^2) and higher-order moments [8]. In general, a statistical signal is completely defined by its pdf and in turn this pdf is uniquely related to the moments. In some cases the entire pdf can be generated using only μ and σ^2, as in the case of a Gaussian distribution, where the pdf of a Gaussian random variable x is given by

$$f_{\mathbf{x}}(x) = \frac{1}{\sqrt{2\pi}\sigma} e^{-(x-\mu)^2/2\sigma^2}.$$

The pdf does not have to be *static*; it could vary with time, making the signal *non-stationary*.

1.2.3 Periodic Signals

A signal $x(t)$ is said to be periodic if

$$x(t) = x(t + nT) \text{ where } T \text{ is a real number and } n \text{ is an integer} \tag{1.3}$$

The periodicity of $x(t)$ is min(nT), which is T. If we rewrite this using the notation $t = k\,\delta t$, where k is an integer and δt is a real number,[3] then

$$x(k\,\delta t) = x\left(k\,\delta t + n\left[\frac{T}{\delta t}\right]\delta t\right) \tag{1.4}$$

$$= x\left(k\,\delta t + n[\hat{N} + \epsilon]\,\delta t\right). \tag{1.5}$$

We write the quantity $T/\delta t = \hat{N} + \epsilon$, where ϵ is a real number[4] less than 1 and $\hat{N}$ is an integer. The practising engineer's *periodicity* in a loose sense is $\hat{N}$; in a strict mathematical sense the periodicity min$\left[n(\hat{N} + \epsilon)\right] = N$ *must* be an integer. N could be very different from $\hat{N}$ or may not exist. This illustrates that periodicity is not preserved while moving from the continuous domain to the discrete domain in all cases in a *strict* sense.

---

[3]This quantity is known as *sampling time*.

[4]A proper choice of δt can make this quantity ϵ almost zero or zero.

1.3 Transformations

The sole aim of this section is to introduce two common signal transformations. A detailed discussion is available in later chapters of this book. Signals are manipulated by performing various mathematical operations for better understanding, presentation and visibility of the signal.

1.3.1 Laplace and Fourier Transforms

One such transformation is the Laplace transform (LT). If $h(t)$ is a time-varying signal, which is a function of time t, then the Laplace transform[5] of $h(t)$ is denoted as $H(s)$, where s is a complex variable, and is defined as

$$H(s) \triangleq \int_{-\infty}^{\infty} h(t)e^{-st}\, \mathrm{d}t. \qquad (1.6)$$

The complex variable s is often denoted as $\sigma + j\omega$, where $j = \sqrt{-1}$. σ is the real part of s and represents the rate of decay or rise of the signal. ω is the imaginary part of s and represents the periodic variation or *frequency* of the signal. The variable s is also sometimes called the complex frequency.

When we are interested only in the frequency content of the signal $h(t)$, we use the Fourier transform, which is denoted[6] $H(\omega)$ and given by

$$H(\omega) \triangleq \int_{-\infty}^{\infty} h(t)e^{-j\omega t}\, \mathrm{d}t. \qquad (1.7)$$

In fact, combining the above equation with the Euler equation,[7] we can derive the *Fourier series*, which is a fundamental transformation for periodic functions.

These transformations are linear in nature, in the sense that the Laplace or Fourier transform of the sum of two signals is the sum of their transforms, and the Laplace or Fourier transform of a scaled version of a signal by some time-independent scale factor is the scaled version of its transform by the same scale factor.

1.3.2 The z-Transform and the Discrete Fourier Transform

When we move from the continuous-time domain to the discrete-time domain, integrands get mapped to summations. Replacing e^s with z and $h(t)$ with h_k in (1.6), we get a new transform in the discrete-time domain known as the z-transform, one

---

[5]This is a two-sided Laplace transform.

[6]We could write it as $H(\omega)$ or $H(j\omega)$ and there is no loss of generality.

[7]Euler equation $e^{j\omega t} = \cos \omega t + j \sin \omega t$.

of the powerful tools representing discrete-time systems. The z-transform of a discrete-time signal h_k is denoted as $H(z)$ and is given by

$$H(z) \triangleq \sum_{k=-\infty}^{\infty} h_k z^{-k}. \tag{1.8}$$

As a simple illustration, consider a sequence of numbers

$$h_k = \begin{cases} 0.5^k, & k \geq 0, \\ 0, & k < 0. \end{cases} \tag{1.9}$$

Using (1.8) for the sequence h_k we get

$$\begin{aligned}
H(z) &= \sum_{k=0}^{\infty} 0.5^k z^{-k} \\
&= \sum_{k=0}^{\infty} (0.5\, z^{-1})^k \\
&= 1 + 0.5\, z^{-1} + 0.25\, z^{-2} + 0.125\, z^{-3} + \cdots. \tag{1.10}
\end{aligned}$$

Using the simple *geometric progression* relation

$$\sum_{k=0}^{\infty} a^k = \frac{1}{1-a}, \quad |a| < 1,$$

we get

$$H(z) = \frac{1}{(1 - 0.5\, z^{-1})}, \tag{1.11}$$

under the condition $\left|0.5 z^{-1}\right| < 1$, that is, $|z| > 0.5$. Note that this condition represents the region of the complex z-plane in which the series (1.10) converges and (1.11) holds, and is called the *region of convergence* (ROC). We can also write h_k in the form

$$h_k = 0.5\, h_{k-1} + \delta_k, \tag{1.12}$$

where δ_k is the *Kronecker delta function*, given by

$$\delta_k = \begin{cases} 1, & k = 0, \\ 0, & k \neq 0. \end{cases} \tag{1.13}$$

If, instead of the variable z, we use the variable $e^{j\Omega}$, where Ω is the frequency, then we get the discrete fourier transform (DFT), which is defined as

$$H(e^{j\Omega}) \triangleq \sum_{k=-\infty}^{\infty} h_k e^{-j\Omega k}. \tag{1.14}$$

The z-transform is thus a discrete-time version of the Laplace transform and the DFT is a discrete-time version of the Fourier transform.

1.3.3 An Interesting Note

We have started with an *infinite* sequence h_k (1.9) which is also represented in the form $H(z)$, known as the *transfer function* (TF), in (1.11) and in a recursive way in (1.12). Thus we have multiple representations of the same signal. These representations are of prime importance and constitute the heart of learning DSP; greater details are provided in later chapters.

1.4   Signal Characterisation

The signal y_k of (1.2) could be any of the signals in Figures 1.1, 1.2 or 1.3 and it could be characterised in the frequency domain by the *non-parametric* spectrum. This spectrum is popularly known as the *Fourier spectrum* and is well understood by everyone. The same signal can also be represented by its *parametric* spectrum, computed from the parameters of the system where the signal originates.

Signals *per se* do not originate on their own; generally a signal is the output of a process defined by its dynamics, which could be linear, non-linear, logical or a combination. Most of the time, due to our limitations in handling the information, we assume a signal to be the output of a linear system, either represented in the continuous-time domain or the discrete-time domain. In this book, we always refer to discrete-time systems. In computing the parametric spectrum, we aim to find the system which has an output of a given signal, say y_k. This constitutes a separate subject known as parameter estimation or system identification.

1.4.1 Non-parametric Spectrum or Fourier Spectrum

If the signal y_k of (1.2) is known for $k = 1, 2, \ldots, N$, it can also be represented as a vector

$$\mathbf{Y}_N = [y_1 \cdots y_N]^T, \tag{1.15}$$

where $(\cdot)^T$ denotes transpose, which can be transformed into the frequency domain by the DFT relation

$$g_k = \frac{1}{N} \left\{ \sum_{n=0}^{N-1} (W^{nk} y_{n+1}) \right\}, \tag{1.16}$$

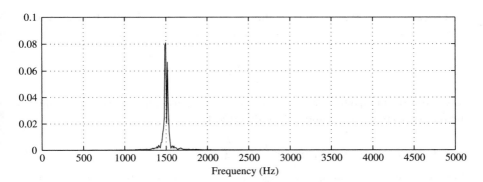

Figure 1.5 Fourier power spectrum $s(k)(1.17)$

where $W = e^{-j2\pi/N}$. In general g_k is a complex quantity. However, we restrict our interest to the *power spectrum* of the signal y_k, a real non-negative quantity given by

$$s_k = g_k g_k^* = |g_k|^2, \qquad (1.17)$$

where $(\cdot)^*$ represents the complex conjugate and $|\cdot|$ represents the absolute value (Figure 1.5). The series S_N, given by

$$\mathbf{S}_N = \{s_1, \ldots, s_N\}, \qquad (1.18)$$

is another way of representing the signal. Even though the signal in the power spectral domain is very convenient to handle, other considerations such as resolution, limit its use for online application. Also, in this representation, the phase information of the signal is lost.

To preserve the total information, the complex quantity g_k is represented as an ordered pair of time series, one representing the *in-phase* component and the other representing the *quadrature* component:

$$\{g_k\} = \{g_k^i\} + j\{g_k^q\}.$$

1.4.2 Parametric Representation

In a parametric representation the signal y_k is modelled as the output of a linear system which may be an all-pole or a pole–zero system [2]. The input is assumed to be white noise, but this input to the system is not accessible and is only conceptual in nature. Let the system under consideration be

$$x_k = \sum_{i=1}^{p} a_i x_{k-i} + \sum_{j=0}^{q-1} b_{j+1}\gamma_{k-j}, \qquad (1.19)$$

$$y_k = x_k + \text{noise}, \qquad (1.20)$$

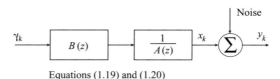

Equations (1.19) and (1.20)

where γ_k is white noise and y_k is the output of the system. Equation (1.19) can also be represented as a transfer function (TF) [3, 4]:

$$H(z) = \frac{B(z)}{A(z)} \tag{1.21}$$

or

$$X(z) = \left\{ \frac{B(z)}{A(z)} \right\} \Gamma(z) \tag{1.22}$$

or in the delay operator notation[8]

$$x_k = \left\{ \frac{B(z)}{A(z)} \right\} \gamma_k, \tag{1.23}$$

where $X(z)$ and $\Gamma(z)$ are the z-transforms of x_k and γ_k, respectively, and

$$A(z) = 1 - \sum_{i=1}^{p} a_i z^{-i}, \tag{1.24}$$

$$B(z) = \sum_{j=0}^{q-1} b_{j+1} z^{-j}, \tag{1.25}$$

are the z-transforms of $\{1, -a_1, -a_2, \ldots, -a_p\}$ and $\{b_1, b_2, \ldots, b_q\}$, respectively. We define the parameter vector as

$$\mathbf{p} = [a_1, a_2, \ldots, a_p, b_1, b_2, \ldots, b_q]^T. \tag{1.26}$$

The parameter vector $\mathbf{p}$ completely characterises the signal x_k. The system defined by (1.19) is also known as an autoregressive moving average (ARMA) model [1, 6]. Note that the TF $H(z)$ of this model is a rational function of z^{-1}, with p poles and q zeros in terms of z^{-1}. An ARMA model or system with p poles and q zeros is conventionally written ARMA (p, q).

---

[8]Some authors prefer to use the delay operator and the complex variable z^{-1} interchangeably for convenience. In the representation $x_{k-1} = z^{-1} x_k$ here, we have to treat z^{-1} as a delay operator and not as a complex variable. In a loose sense, we can exchange the complex variable and the delay operator without much loss of generality.

1.4.2.1 Parametric Spectrum

Given the parameter vector **p** (1.26) we obtain the *parametric spectrum* $|H(e^{j\omega})|$ as

$$|H(e^{j\omega})| = \left|\frac{B(e^{j\omega})}{A(e^{j\omega})}\right| \tag{1.27}$$

$$= \left|\frac{\sum_{m=0}^{q} b_{m+1} e^{-j\omega m}}{1 - \sum_{n=1}^{p} a_n e^{-j\omega n}}\right|. \tag{1.28}$$

Using the given value of **p**, we obtain the parametric spectrum via (1.28). The function $|H(e^{j\omega})|$ is periodic (with period 2π) in ω.

When the parameter vector **p** is real-valued, $|H(e^{j\omega})|$ is also symmetric in ω, and we need to evaluate (1.28) only for $0 < \omega < \pi$. Figure 1.6 is the parametric spectrum of the narrowband signal. We have obtained this spectrum by first obtaining the parameter vector **p** (1.26) of the given time series and substituting this value in (1.28).

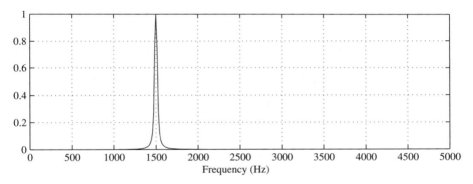

Figure 1.6   Parametric spectrum $\|H(e^{jw})\|$ using (1.28)

1.5   Converting Analogue Signals to Digital

An analogue signal $y(t)$ is in reality a mapping function from the real line, to the real line, defined as $\mathcal{R} \to \mathcal{R}$, where $\mathcal{R}$ denotes the set of real numbers. When converted to a digital signal, this function goes through transformations and becomes modified into another signal which mostly preserves the information, depending on how the conversion is done. The process of modification or morphing has three distinct stages:

- Windowing
- Digitisation: sampling
- Digitisation: quantisatioin

1.5.1 Windowing

The original signal could be very long and non-periodic, but due to physical limitations we observe the signal only for a finite duration. This results in multiplication of the signal by a rectangular window function $R(t)$, giving an observed signal

$$\tilde{y}(t) = y(t)R(t), \tag{1.29}$$

where

$$R(t) = \begin{cases} 1, & 0 \le t < T_w, \\ 0, & \text{otherwise}. \end{cases} \tag{1.30}$$

T_w is the observation time interval. In general, $R(t)$ can take many forms and these functions are known as windowing functions [5].

1.5.2 Sampling

In reality, a band-limited analogue signal $y(t)$ needs to be sampled, resulting in a discrete-time signal $\{y_k\}$, converting the function as a mapping from the integer line to the real line $\mathcal{I} \to \mathcal{R}$, where $\mathcal{I}$ denotes the set of integers. We express $\{y_k\}$ as

$$\{y_k\} = \sum_{k=-\infty}^{\infty} y(t)R(t)\delta(t - kT_s)$$

$$= \left[\sum_{k=-\infty}^{\infty} y(t)\delta(t - kT_s) \right] R(t). \tag{1.31}$$

where $\delta(t)$ is the unit impulse (Dirac delta function)

$$\delta(t) = \begin{cases} 1, & t = 0, \\ 0, & t \ne 0. \end{cases} \tag{1.32}$$

T_s is the sampling time. The sampling process is defined via (1.31) and is shown in Figure 1.7(a). This process generates a sampled or discrete signal. The Fourier transform of the sampled signal y_k (Figure 1.7(b)) is computed and shown in Figure 1.8.

The striking feature in Figure 1.8 is the periodic replication of the narrowband spectrum. Revisiting Fourier series will help us to understand the effect of sampling. The Fourier series is indeed a discrete spectrum or a line spectrum, and results from the periodic nature of a signal in the time domain. Any transformation from the time domain to the frequency domain maps periodicity

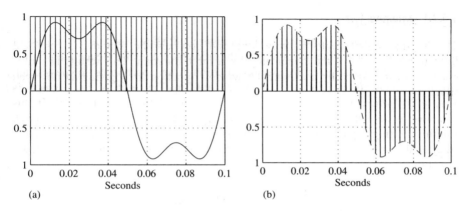

Figure 1.7   Discrete signal generation via (1.31)

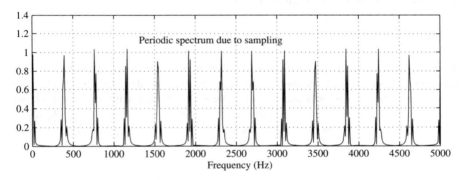

Figure 1.8   Spectrum of sampled signal y_k

into sampling, and vice versa. What it means is that sampling in the time domain brings periodicity in the frequency domain, and vice versa, as follows.

Time domain	Frequency domain
Sampling	Periodicity
Periodicity	Sampling

In fact, the Poisson equation depicts this as

$$S_y(f) = \sum_{n=-\infty}^{\infty} S_y\left(f - \frac{n}{T_s}\right),\tag{1.33}$$

where $S_y(\cdot)$ is the *Fourier spectrum* of the signal $y(t)$, which is periodic in f with period $\frac{1}{T_s}$, shown in Figure 1.8.

1.5.2.1 Aliasing Error

A careful inspection of (Equation 1.33 and Figure 1.8) shows there is no real loss of information except for the periodicity in the frequency domain. Hence the original signal can be reconstructed by passing it through an appropriate lowpass filter. However, there is a condition in which information loss occurs, which is $1/T_s \geq 2f_c$, where f_c is the cut-off frequency of the band-limited signal. If this condition is not satisfied, a wraparound occurs and frequencies are not preserved. But this aliasing is put to best use for downconverting the signals without using any additional hardware, like mixers in digital receivers, where the signals are bandpass in nature.

1.5.3 Quantiaation

In addition, the signal gets quantised due to finite precision analogue-to-digital converters. The signal can be modelled as

$$y_k = \lfloor y_k \rfloor + \nu_k, \tag{1.34}$$

where $\lfloor y_k \rfloor$ is a finite quantised number and ν_k is a *uniformly distributed* (UD) random number of $\frac{1}{2}$ LSB. Sampled signal y_k and $\lfloor y_k \rfloor$ are depicted in Figure 1.9(a) and the quantisation error is shown in Figure 1.9(b). In this numerical example we have used 10 (±5) levels of quantisation, giving an error (γ_k) between $\pm\frac{1}{10}$, which can be seen in Figure 1.9(b). The process of moving[9] signal from one domain to the

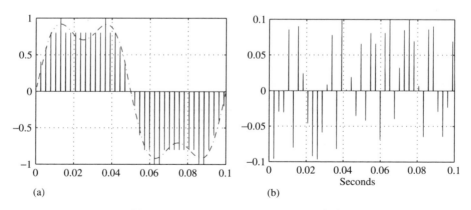

(a)   (b)

Figure 1.9   Quantised discrete signal $\lfloor y_k \rfloor$

---

[9]This movement is because we want to do digital signal processing.

other domain is as follows:

$$y(t) \quad \rightarrow \quad y_k \quad \rightarrow \quad \lfloor y_k \rfloor \quad \rightarrow \quad \{\hat{y}_k\}$$

| Continous | Discrete | Quantised Discrete | Windowed Quantised Discrete |

In reality we get only *windowed, discrete* and *quantised* signal $\{\lfloor y_k \rfloor\} = \{\hat{y}_k\}$ at the processor. Figure 1.10 depicts a 4-bit quantised, discrete and rectangular windowed signal.

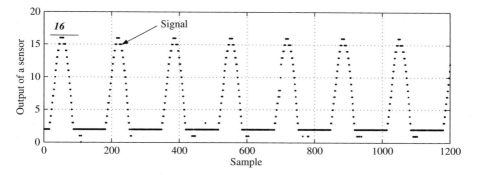

Figure 1.10 Windowed discrete quantised signal

1.5.4 Noise Power

The sample signal y_k on passing through a quantiser such as an analogue-to-digital (A/D) converter results in a signal $\lfloor y_k \rfloor$, and this is shown in Figure 1.9 along with the quantisation noise. The quantisation noise $\nu_k = y_k - \lfloor y_k \rfloor$ is a random variable (rv) with uniform distribution. The variance σ^2 is given as $(1/12)(2^{-n})^2$, where n is the length of the quantiser in bits. The noise power in decibels (dB) is given as $10 \log \sigma^2 = -10 \log(12) - [20 \log(2)]n$. If we assume that the signal y_k is equiprobable along the range of the quantiser, it becomes a uniformly distributed signal. We can assume without any loss of generality the range as 0 to 1. Then the signal power is $10 \log(12)$. The signal-to-noise ratio (SNR) is $20n \log(2)$ dB or 6 dB per bit.

Mathematically, the given original signal got corrupted due to the process of sampling and converting into physical world, real numbers. These are the theoretical modifications *alone*.

1.6   Signal Seen by the Computing Engine

With so much compulsive morphing, the final digitised signal[10] takes the form

$$\{y_k\} = \left\lfloor \left(\sum_{k=-\infty}^{\infty} y(t)\delta(t - T_s k) \right) R(t) \right\rfloor + \nu_k. \tag{1.35}$$

$$y_k = \hat{y}_k + \nu_k. \text{ modelling of } Discrete \text{ Signal}$$

In (1.35) we have not included any noise that gets added due to the channel through which the signal is transmitted. Besides that, in reality the impulse function $\delta(t)$ is approximated by a narrow rectangular pulse of finite duration.

1.6.1 Mitigating the Problems

Quantisation noise is purely controlled by the number of bits of the converter. The windowing effect is minimised by the right choice of window functions or choosing data of sufficient duration. There are scores of window functions [5] to suit different applications.

1.6.1.1 Anti-Aliasing Filter

It is necessary to have a band-limited signal before converting the signal to discrete form, and invariably an analogue lowpass filter precedes an A/D converter, making the signal band-limited. The choice of the filter depends on the upper bound of the sampling frequency and on the nature of the signal.

1.6.2 Anatomy of a Converter

There are three distinct parts to an A/D converter, as shown in Figure 1.11:

- Switch
- Analogue memory
- Conversion unit

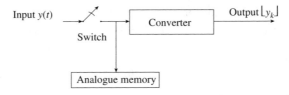

Figure 1.11   Model of an A/D converter

---

[10] The signal $\hat{y}_k$ is a floating-point representation of the signal $\lfloor y_k \rfloor$.

The switch is controlled by an extremely accurate timing device. When it is switched on it facilitates the transfer of the input signal $y(t)$ at that instant to an analogue memory; after transfer, it is switched off, holding the value, and conversion is initiated. In specifying an A/D device there are four important timings: on time t_{on} of the switch is a non-zero quantity, because any switch takes a finite time to close; hold time t_{hold} is the period when the signal voltage is transferred to analogue memory; off time t_{off} is also non-zero, because a switch takes a finite time to open; finally, there is the conversion time t_c.

The sampling time T_s is the sum of all these times, thus $T_s = t_{on} + t_{hold} + t_{off} + t_c$. As technology is progressing, the timings are shrinking, and with good hardware architecture, sampling speeds are approaching 10 gigasamples per second with excellent resolution (24-bit). Way back in 1977 all these units were available as independent hardware blocks. Nowadays, manufacturers are also including *anti-aliasing* filters in A/D converters, apart from providing as a single chip.

There are many conversion methods. The most popular ones are the *successive approximation* method and a flash conversion method based on parallel conversion, using high slew rate operational amplifiers. A wide range of these devices are available in the commercial market.

1.6.3 The Need for Normalised Frequency

We enter the discrete domain once we pass through an A/D converter. We have *only numbers and nothing but numbers*. In the discrete-time domain, only the normalised frequency is used. This is because, once sampling is performed and the signal is converted into numbers, the real frequencies are no longer of any importance. If f_{actual} is the actual frequency, f_n is the normalised frequency, and f_s is the sampling frequency, then they are related by

$$f_{actual} = f_n \times f_s. \tag{1.36}$$

Due to the periodic and symmetric [2] nature of the spectrum, we need to consider f_n only in the range $0 < f_n < 0.5$. This essentially follows from the Nyquist theorem, where the minimum sampling rate is twice the maximum frequency content in the signal.

1.6.4 Care before Sampling

Once the conversion is over, damage is either done or not done; there are no halfway houses. Information is preserved or not preserved. All the care must be taken before conversion.

1.7 It Is Only Numbers

Once the signal is converted into a set of finite accurate numbers that can be represented in a finite-state machine, it is only a matter of playing with numbers.

Figure 1.12   It is playing with numbers

There are well-defined, logical, mathematical or heuristic rules to play with them. To demonstrate this concept of playing (Figure 1.12), we shall conclude this chapter with a simple example of playing with numbers.

Let $\{y_k\} = \{y_1, y_2, \ldots\}$ be a sequence of numbers so acquired. Its mean at *discrete* time k is defined as

$$\mu_k = \frac{1}{k} \sum_{i=1}^{k} y_i. \tag{1.37}$$

The above way of computing the mean is known as a *moving-average* process. We recast the (1.37) as $k\mu_k = \sum_{i=1}^{k} y_i$ and for $k + 1$ we write it as

$$(k + 1)\mu_{k+1} = \sum_{i=1}^{k+1} y_i$$

$$= \sum_{i=1}^{k} y_i + y_{k+1} \tag{1.38}$$

$$= k\mu_k + y_{k+1}. \tag{1.39}$$

We can rewrite (1.39) as

$$\mu_{k+1} = \left(\frac{k}{k+1}\right)\mu_k + \left(\frac{1}{k+1}\right)y_{k+1} \tag{1.40}$$

$$= a_1(k)\mu_k + b_1(k)y_{k+1}. \tag{1.41}$$

It can easily be seen that (1.41) is a *recursive* method for computing the mean of a given sequence. Equation (1.41) is also called an *autoregressive* (AR) process.[11] It will be shown in a later chapter that it represents a lowpass filter.

1.7.1 Numerical Methods

The numerical methods existing today for solving differential equations, they all convert them into *digital filters* [5, 7]. Consider the simplest first-order differential equation

$$\dot{x} = -0.5x + u(t), \tag{1.42}$$

where $u(t)$ is finite for $t \geq 0$.

We realise that slope $\dot{x}$ can be approximated[12] as $T^{-1}(x_{k+1} - x_k)$, where T is the step size, and substituting this in (1.42) gives

$$x_{k+1} = (1 - 0.5\,T)x_k + Tu_k, \tag{1.43}$$

which is the same as

$$x_k = 0.95\,x_{k-1} + 0.1\,u_{k-1} \tag{1.44}$$

for a step size of 0.1. It is very important to note that (1.12), (1.41) and (1.44) come from different backgrounds but they have the same structure as (1.19). In (1.44) the step size, which is equivalent to the sampling interval, is an important factor in solving the equations and also has an effect on the coefficients. The performance of the digital signal processing (DSP) algorithms depends quite a lot on the sampling rate. It is here that common sense fails, when people assume that sampling at a higher rate works better. Generally, most DSP algorithms work best when sampling is four times or twice the Nyquist frequency.

1.8 Summary

In this chapter we discussed the types of signals and the transformations a signal goes through before it is in a form suitable for input to a DSP processor. We

---

[11]Refer to (1.19) in generic form.

[12]*Euler* method.

introduced some common transforms, time and frequency domain interchanges, and the concepts of windowing, sampling and quantisation plus a model of an A/D converter. The last few sections give a flavour of DSP.

References

1. M. I. Skolnik, *Introduction to Radar Systems*. New York: McGraw-Hill, 2002.
2. S. M. Kay, *Modern Spectral Estimation Theory and Applications*. Englewood Cliffs NJ: Prentice Hall, 1988.
3. P. Young, *Recursive Estimation and Time Series Analysis*, p. 266. New York: Springer-Verlag, 1984.
4. P. Young, *Recursive Estimation and Time Series Analysis*, Ch. 6, p. 108. New York: Springer-Verlag, 1984.
5. R. W. Hamming, *Digital Filters*, pp. 4–5. Englewood Cliffs NJ: Prentice Hall, 1980.
6. J. M. Mendel, *Lessons in Digital Estimation Theory*. Englewood Cliffs NJ: Prentice Hall, 1987.
7. R. D. Strum and D. E. Kirk, *First Principles of Discrete Systems and Digital Signal Processing*. Reading MA: Addison-Wesley, 1989.
8. A. Papoulis, *Probability and Statistics*. Englewood Cliffs NJ: Prentice Hall, 1990.

2

Revisiting the Basics

In this chapter we revisit the basics, covering linear systems, some important relevant concepts in number systems, basics of optimisation [1–3], fundamentals of random variables, and general engineering background with illustrations. Algebra is the best language by which an engineer can communicate with precision and accuracy, hence we have used it extensively where normal communication is difficult.

A good understanding of linear system [4] is the basic foundation of digital signal processing. In reality only very few systems are linear and most of the systems we encounter in real life are non-linear. Human beings are non-linear and their behaviour cannot be predicted accurately based on past actions. In spite of this, it is essential to thoroughly understand linear systems, with an intention to approximate a complex system by a piecewise linear model. Sometimes we could also have short-duration linear systems, in a temporal sense.

2.1 Linearity

The underlying principles of linear systems are *superposition* and *scaling*. As an abstraction, the behaviour of the system when two input conditions occur simultaneously is the same as the sum of the outputs if the two occur separately. Mathematically, for example, consider a function $f(x)$ whose values at x_1 and x_2 are $f(x_1)$ and $f(x_2)$, then

$$f(x_1 + x_2) = f(x_1) + f(x_2). \qquad (2.1)$$

This is the law of superposition. The *scaling* property demands a proportionality feature from the system. If the function value at ax is $f(ax)$, then this property demands

$$af(x) = f(ax). \qquad (2.2)$$

Digital Signal Processing: A Practitioner's Approach   K. V. Rangarao and R. K. Mallik
© 2005 John Wiley & Sons, Ltd

This is the law of scaling. We see that a simple function, which looks apparently linear, *fails* to adhere to these two basic rules. Consider $f(x) = 2x + 5$. We have $f(2) = 9$ and $f(5) = 15$ but $f(5 + 2) = 19$, which is not the same as $f(2) + f(5)$ since $19 \neq 24$. Let us look at $f(3 \times 2) = f(6) = 17$ and $3 \times f(2) = 27$. This function fails to obey both the rules, hence it is *not* linear. We can modify this function by a simple transformation $g(x) = f(x - \frac{5}{2})$ and make it linear.

2.1.1 Linear Systems

Consider a single-input single-output (SISO) system that can be excited by a sequence u_k resulting in an output sequence y_k. It is depicted in Figure 2.1. The

Figure 2.1   A single-input single-output system

sequence u_k could be finite while y_k could be *infinite*, which is common. For the system to be linear,

$$
\begin{aligned}
u_k^1 &\Longrightarrow y_k^1, \\
u_k^2 &\Longrightarrow y_k^2, \\
(u_k^1 + u_k^2) &\Longrightarrow y_k^3, \\
(au_k^1) &\Longrightarrow y_k^4.
\end{aligned}
\tag{2.3}
$$

Thus, we must have $y_k^3 = y_k^1 + y_k^2$ and $y_k^4 = ay_k^1$, where a is an arbitrary constant.

Consider two cascaded linear systems A and B with input u_k and output y_k. Linearity demands that exchanging systems will not alter the output for the same input. This is indicated in Figure 2.2.

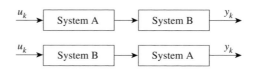

Figure 2.2   Commutative property

2.1.2 Sinusoidal Inputs

Under steady state, linear systems display wonderful properties for sinusoidal inputs. If $u_k = \cos(2\pi f k)$ then the output is also a sinusoidal signal given as $y_k = A\cos(2\pi f k + \phi)$. This output also can be written as $y_k = y_k^i + y_k^q$, where $y_k^i = A_1 \cos(2\pi f k)$ and $y_k^q = B_1 \sin(2\pi f k)$. Mathematicians represent the sequence

y_k as a complex sequence $y_k = y_k^i + jy_k^q$, where $j = \sqrt{-1}$. Engineers treat y_k as an *ordered pair* of sequences, popularly known as *in-phase* and *quadrature* sequences. Note that there is no change in frequency but only change in *amplitude* and *phase*. This concept is used in many practical systems, such as network analysers, spectrum analysers and transfer function analysers.

2.1.3 Stability

Another important property defining the system is its stability. By definition if $u_k = \delta_k$ and the corresponding output is $y_k = h_k$, and if the sequence h_k is convergent, then the system is stable. Here δ_k is the unit sample.[1] We can also define a summation $S = \sum_{k=0}^{\infty} h_k$ and if S is finite, then the system is stable. The physical meaning is that the area under the discrete curve h_k must be finite. There are many methods available for testing convergence. One popular test is known as the *ratio test*, which states that if $|h_{k+1}/h_k| < 1$, then the sequence h_k is convergent, hence the system is stable.

2.1.4 Shift Invariance

Shift invariance is another important desirable property. A shift in the input results in an equal shift in the output ($u_{k-N} \implies y_{k-N}$). This property is known as shift invariance. Linear and shift invariant (LSI) systems have very interesting properties and are mathematically tractable. This is the reason why many theoretical derivations assume this property.

2.1.5 Impulse Response

If the input u_k is a unit sample δ_k, then the system response y_k is called a *unit sample response* of the system, h_k. This has a special significance and completely characterises the system. This sequence could be finite or infinite. The sequence h_k is said to be finite if

$$h_k = \text{finite for } k < N$$
$$= 0 \text{ for } k \geq N. \tag{2.4}$$

If the impulse response h_k is finite as defined in (2.4), then the system is a finite impulse response (FIR) system and it has a special significance. Figure 2.3 shows a typical unit sample response of a system defined by (2.13) and the unit sample response for this system is infinite. Traditionally, systems in the discrete domain originate from differential equations describing a given system, and these differential equations take the form of *difference* equations in the discrete domain.

---

[1] Impulse response is used in continous systems.

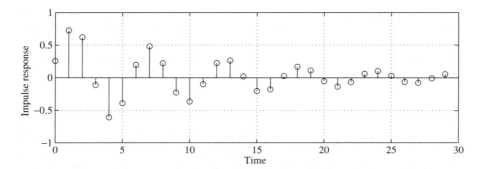

Figure 2.3 Impulse response h_k

2.1.5.1 Convolution: Response to a Sequence

Consider a system defined by its unit sample response h_k, and let the input sequence be u_k of length N. Then the response due to u_1 can be obtained using the scaling property as $u_1 h_k$, and the response due to $u_2, u_3, u_4, \ldots, u_{N+1}$ can be obtained using the scaling and shifting properties as

$$u_2 h_{k-1}, u_3 h_{k-2}, u_4 h_{k-3}, \ldots, u_{N+1} h_{k-N}.$$

Now, using superposition we obtain the response due to the sequence u_k by summing all the terms as

$$u_1 h_k + u_2 h_{k-1} + u_3 h_{k-2} + u_4 h_{k-3} + \cdots + u_{N+1} h_{k-N}. \tag{2.5}$$

We need to remember that the term $\{u_1 h_k\}$ is a sequence obtained by multiplying a scalar u_1 with another sequence h_k. This mental picture is a **must** for better understanding the operations below. We obtain the response y_k through a familiar operation known as convolution:[2]

$$y_k = \sum_{n=0}^{n=N} u_{n+1} h_{k-n}. \tag{2.6}$$

While deriving (2.6) we have used shift invariance properties. The above convolution operation is represented symbolically as $y_k = u_k \otimes h_k$. The binary operator $\otimes$ which represents convolution is linear and follows all the associative and commutative properties. In the frequency domain, convolution takes the form of multiplication, $Y(j\omega) = U(j\omega) \times H(j\omega)$, where all the $j\omega$ functions are corresponding frequency domain functions.

---

[2] $y(t) = \int_{-\infty}^{\infty} u(\tau) h(t - \tau) \, d\tau.$

Convolution is also known as filtering and it is also equivalent to multiplying two polynomials, $H(z) \times U(z)$. If we have a sequence h_k of length m and a sequence u_k of length N, then we can see that the output sequence is of size $m + N$.

Consider the computational complexity of (2.6). Computing $u_1 h_k$ needs m multiplications and m additions. Assuming there is not much difference between addition and multiplication, in the floating-point domain, the computational burden is $2m$ operations for one term, and the whole of (2.6) requires $(2m \times N + N - 1)$ operations. For $N = 100, m = 50$, which are typical values, we need to perform 10 099 operations. As an engineering approximation, we can estimate this as $2mN$ operations.

Computing the convolution of two sequences using (2.6) is computationally expensive. Less expensive methods are available based on discrete Fourier transforms (DFTs). This is because there are fast methods for computing DFTs.

2.1.5.2 Autocorrelation Function

Another important operation between two sequences, say u_k and h_k, is correlation. Typically u_k is an input sequence to a linear system and h_k is the impulse response of the same system. The correlation function is defined as

$$r_k^{uh} = \sum_{n=0}^{n=N} u_{n+1} h_{k+n}. \tag{2.7}$$

This is also written using the convolution operator as $y_k = u_k \otimes h^*_{-k}$, where h^*_{-k} is the conjugate symmetric part of h_k. When both sequences in (2.7) are the same and equal to the impulse response h_k then r_k^{uh} becomes the autocorrelation[3] function r_k of the system, shown in Figure 2.4(a). The autocorrelation function r_k has special

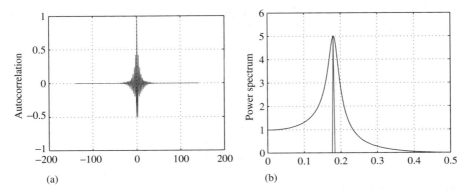

Figure 2.4 Relation between autocorrelation r_k and power spectrum $\|H(j\omega)\|$

---

[3] $r(t) = \int_{-\infty}^{\infty} h(\tau) h(t - \tau) \, d\tau.$

significance in linear systems and is given as

$$r_k = \sum_{n=0}^{n=N} h_{n+1} h_{k+n} = h_k \otimes h_{-k}^*. \tag{2.8}$$

The DFT of r_k results in the power spectrum of the system. The DFT of the given autocorrelation function r_k of a linear system is shown in Figure 2.4(b). Note that r_k of a linear system and its power spectrum $|H(e^{j\omega})|$ are synonymous and can be used interchangeably. In fact, the Wiener–Khinchin theorem states the same.

2.1.6 Decomposing h_k

The impulse response of a given linear system can be considered as a sum of impulse responses of groups of first- and second-order systems. This is because the system transfer function $H(z) = B(z)/A(z)$ can be decomposed into partial fractions of first order (complex or real) or a sum of partial fractions of all-real second-order and first-order systems. This results in the given impulse response h_k being expressed as a sum of impulse responses of several basic responses:

$$h_k = \sum_{i=1}^{M} \alpha_i h_k^i \quad \text{where } \alpha_i \text{ is a constant,} \tag{2.9}$$

where h_k^i is the unit sample response of either a first-order system or a second-order system.

2.2 Linear System Representation

We would like to avoid continuous systems in this textbook, but we are compelled to mention them, although this is probably the last time. Consider the popular second-order system

$$\ddot{x} = -2\delta\omega_n \dot{x} - \omega_n^2 x + \omega_n^2 u(t)$$
$$= -4\pi\delta f_n \dot{x} - 4\pi^2 f_n^2 x + 4\pi^2 f_n^2 u(t). \tag{2.10}$$

Here

$$\ddot{x} = \frac{d^2 x}{dt^2} \quad \text{and} \quad \dot{x} = \frac{dx}{dt}.$$

Taking numerical $\delta = 0.1$ and $f_n = 100\,\text{Hz}$ we get

$$\ddot{x} = -125.7\dot{x} - 394\,784x + 394\,784u(t). \tag{2.11}$$

2.2.1 Continuous to Discrete

Let us take a numerical method [5] known as the Tustin approximation for sampling times of 0.001 s and 0.002 s and apply it to the continuous system characterised by (2.11). We get the difference equations as

$$y_k = 1.5519\, y_{k-1} - 0.8918\, y_{k-2} + 0.08497\, u_k + 0.1699\, u_{k-1} + 0.08497\, u_{k-2}, \quad (2.12)$$
$$y_k = 0.7961\, y_{k-1} - 0.8347\, y_{k-2} + 0.2596\, u_k + 0.5193\, u_{k-1} + 0.2596\, u_{k-2}. \quad (2.13)$$

Note that we have two different difference equations representing the same physical system; this is due to a change in sampling rate. Equation (2.13) has two components: an autoregressive (feedback) component and a moving-average (feedforward) component. By using the notation $y_{k-1} = z^{-1} y_k$, we define vectors $\mathbf{y}_z$ and $\mathbf{u}_z$:

$$\mathbf{y}_z^t = [1 \quad z^{-1} \quad z^{-2}] y_k \quad \text{and} \quad \mathbf{u}_z^t = [1 \quad z^{-1} \quad z^{-2}] u_k.$$

And we define the coefficient vectors as

$$\mathbf{a}^t = [1 \quad -0.7961 \quad 0.8347] \quad \text{and} \quad \mathbf{b}^t = [0.2596 \quad 0.5193 \quad 0.2596].$$

Using this notation, we write (2.13) in vector form as $\mathbf{a}^t \mathbf{y}_z[y_k] = \mathbf{b}^t \mathbf{u}_z[u_k]$. In this textbook the delay operator z^{-1} and the complex variable z^{-1} have been used interchangeably. We continue to do this even though mathematicians may object. We believe that algebraically there is no loss of generality.

2.2.2 Nomenclature

For historical reasons, specialists from different backgrounds have different names for (2.14) and (2.15). However, they are identical in a strict mathematical sense.

Statisticians use the term Autoregressive (AR) [6]; control engineers call it an all-pole system; and communication engineers call it an infinite impulse response (IIR) filters. The statisticians' moving-average (MA) process is called a feedforward system by control engineers and a finite impulse response (FIR) filter by communication engineers so we have chosen names to suit the context of this textbook; readers may choose other names to suit their own contexts.

2.2.3 Difference Equations

We can write (2.12) or (2.13) in the form

$$\mathbf{a}^t \mathbf{y}_z(y_k) = \mathbf{b}^t \mathbf{u}_z(u_k), \quad (2.14)$$

which can be written as

$$y_k - \sum_{i=1}^{p}(a_{i+1}y_{k-i}) = \sum_{i=1}^{q}(b_i u_{k-i+1}). \qquad (2.15)$$

In (2.15) the coefficient a_1 is always one. Equations (2.14) and (2.15) are the generic linear difference equations of a pth order AR process (left-hand side) and a qth order MA process (right-hand side). There are many technical names for (2.14) and (2.15).

2.2.4 Transfer Function

Recognising (2.14) as the dot product of the vectors, it can be written as

$$A(z) = \mathbf{a}'\mathbf{y}_z \quad \text{and} \quad B(z) = \mathbf{b}'\mathbf{u}_z, \qquad (2.16)$$

where $B(z)$ is called the numerator polynomial and $A(z)$ is called the denominator polynomial. We can recast the equations as $y_k = [B(z)/A(z)]u_k$. The roots of $B(z)$ and $A(z)$ have greater significance in the theory of digital filters, and for understanding physical systems and their behaviour. The impulse response of the system (2.13) is depicted in Figure 2.3; it is real. It is not necessary that the response be a single sequence; it could be an ordered pair of sequences, making it complex. We can write $B(z)/A(z)$ as

$$\begin{aligned} \frac{B(z)}{A(z)} &= \frac{0.2596(1 + 2z^{-1} + z^{-2})}{(1 - 0.7961z^{-1} + 0.8347z^{-2})} \\ &= \frac{0.2596(1 - z^{-1})^2}{(1 - z^{-1}0.9136e^{j1.12})(1 - z^{-1}0.9136e^{-j1.12})}. \end{aligned} \qquad (2.17)$$

The same information is shown in Figure 2.5 as a pole–zero plot. We can decompose it into partial fractions as

$$\frac{B(z)}{A(z)} = 0.2596(1 - z^{-1})^2 \left[\frac{j0.6080z}{(1 - z^{-1}0.9136e^{j1.12})} - \frac{j0.6080z}{(1 - z^{-1}0.9136e^{-j1.12})} \right], \qquad (2.18)$$

as stated in (2.9). However, it is a recommended practice to leave it as a second-order function if the roots are complex. Second-order systems have great significance and the standard representation is given below, where $r = 0.9136$ is the radial pole position at an angle of $\theta = \pm 1.12$ radians in the z-plane.

$$\frac{B(z)}{A(z)} = \frac{0.2596(1 + 2z^{-1} + z^{-2})}{[1 - 2(0.9136)\cos(1.12)z^{-1} + (0.9136)^2 z^{-2}]}. \qquad (2.19)$$

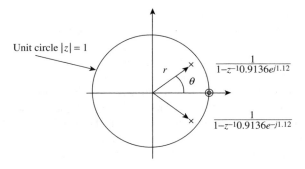

Figure 2.5 Polar plot of the second-order system in (2.17)

2.2.5 Pole–Zero Representation

The polynomials $A(z)$ and $B(z)$ are set equal to zero and solved for z. The roots of the numerator polynomial are called zeros and the roots of the denominator polynomial are called poles of the system. Note that the continuous system represented by (2.10) is defined by only two parameters $[\delta, \omega_n]$ but by converting into discrete form we get three parameters $[0.2596, 0.9136, 1.12]$, as shown in (2.17) and (2.19). The addition of a third parameter is due to the sampling time T_s. The parameter δ represents damping and ω represents the natural angular frequency; they roughly translate into the radial position of the pole and the angular position of the pole, respectively.

2.2.6 Continuous to Discrete Domain

Converting from the continuous domain to the discrete domain also affects the pole positions due to sampling rate. Figure 2.6 depicts the pole positions (r, θ) for

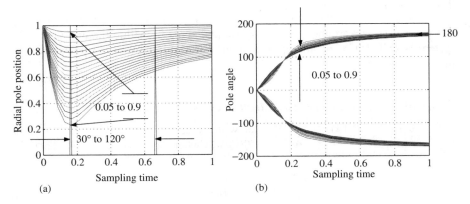

Figure 2.6 Continuous to discrete sensitivities of the system in (2.11)

various values of δ in the range $0.05 < \delta < 0.9$ and for different sampling times. We can use Figure 2.6 to make a number of inferences. In this figure we have shown sampling time as 0 to 1; 0 corresponds to an infinite sampling rate whereas 1 corresponds to the Nyquist sampling rate (twice the maximum frequency). The damping factor δ has been varied in the same range for Figure 2.6(a) and (b). A quick look at Figure 2.6(b) shows that the pole angle is insensitive to variations of δ.

Time has an *approximately* linear relation with pole angle θ. Taking advantage of this relation, we have positioned two markers in Figure 2.6(a) at $30°$ and $120°$, which correspond to 12 samples/cycle and 3 samples/cycle, respectively. The pole plot in Figure 2.6(a) shows that the difference equation numerically behaves well in this window.

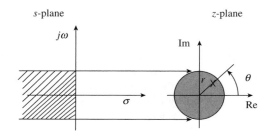

Figure 2.6A   Comparing the s-plane with the z-plane

In discrete systems, roots are presented in polar form rather than rectangular form. This is because the s-plane, which represents the continuous domain, is partitioned into a left half and a right half by the $j\omega$ axis running from ∞ to $-\infty$ (Figure 2.6A). Such an infinite axis is folded into the sum of an infinite number of circles of unit radius in the z-plane, which is the representation in the discrete domain. The top half of the circle (0 to $-\pi$) represents DC to Nyquist frequency and the bottom half (0 to π) represents the negative frequencies.

In fact, periodicity and conjugate symmetry properties can be best understood by this geometric background. If all the poles and zeros lie within the unit circle, the system is called a *minimum phase* system. For good stability, all poles must be inside the unit circle.

2.2.7 State Space Representation

Let us consider (2.14) again. We can write it as $y_k = [B(z)/A(z)]u_k$. Using the linearity property, we can write $y_k = B(z)(w_k)$ where $w_k = [1/A(z)](u_k)$. Then

$$
\begin{aligned}
w_k &= 0.7961w_{k-1} - 0.8347w_{k-2} + u_k, \\
w_{k-1} &= w_{k-1} + 0w_{k-2} + 0u_k, \\
y_k &= 0.2596w_k + 0.5193w_{k-1} + 0.2596w_{k-2}.
\end{aligned}
\tag{2.20}
$$

We can modify the last equation in (2.20) as

$$y_k = 0.2596(0.7961w_{k-1} - 0.8347w_{k-2} + u_k) + 0.5193w_{k-1} + 0.2596w_{k-2}.$$
(2.21)

Now the modified set is

$$
\begin{aligned}
w_k &= 0.7961w_{k-1} - 0.8347w_{k-2} + u_k, \\
w_{k-1} &= w_{k-1} + 0w_{k-2} + 0u_k, \\
y_k &= 0.7260w_{k-1} + 0.0429w_{k-2} + 0.2596u_k.
\end{aligned}
$$
(2.22)

This can be represented in standard **A**, **B**, **C**, **D** matrix form as

$$
\begin{pmatrix} w_k \\ w_{k-1} \end{pmatrix} = \begin{pmatrix} 0.7961 & -0.8347 \\ 1 & 0 \end{pmatrix} \begin{pmatrix} w_{k-1} \\ w_{k-2} \end{pmatrix} + \begin{pmatrix} 1 \\ 0 \end{pmatrix} u_k,
$$
(2.23)

$$
y_k = (0.7260 \quad 0.0429) \begin{pmatrix} w_{k-1} \\ w_{k-2} \end{pmatrix} + 0.2596u_k.
$$
(2.24)

The above difference equation is shown in Figure 2.7, which gives the minimum delay realisation. In brief, $\mathbf{w}_k = \mathbf{A}\mathbf{w}_{k-1} + \mathbf{B}u_k$ and $y_k = \mathbf{C}^t\mathbf{w}_{k-1} + \mathbf{D}u_k$.

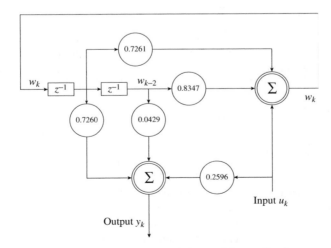

Figure 2.7   Digital filter minimum delay realisation

2.2.8 Solution of Linear Difference Equations

There are many methods available for solving difference equations along the lines of differential equations, by solving for complementary function and particular integral. The advent of powerful software packages such as MATLAB and SIMULINK means that we no longer need to know the details, but if we do understand them, it will help us get the most out of our software.

It is always better to visualise the systems in the form of (2.16) since this best matches the software packages in the commercial market. The transfer function $H(z) = B(z)/A(z)$ can be converted into a sum of second- and/or first-order partial fractions, depending on whether the roots of the polynomial $A(z)$ are real or complex:

$$\frac{B(z)}{A(z)} = \sum(\text{order 2 systems}) + \sum(\text{order 1 systems}). \tag{2.25}$$

A second-order system has a solution of the form $h_k = Ar^{-k}\cos(\theta) + Br^{-k}\sin(\theta)$, whereas a first-order system is of the form $h_k = Cr^{-k}$; the constants can be evaluated from the initial conditions. Equations (2.25) and (2.9) are the same except the domains are different. The overall solution for a specific input u_k can be obtained using (2.6).

2.3 Random Variables

We will introduce two important ideas:

- Continuous random variable (CRV)
- Discrete random variable (DRV)

Note that not every discrete system is a DRV; discrete systems and discrete random variables are separate concepts. Sampling or converting into a discrete signal show its effect on the autocorrelation function r_k but not on its probability density function $f_x(x)$. For completeness, we recollect that a random variable (rv) is a function mapping from a sample space to the real line.

Consider a *roulette wheel* commonly used in casinos. The sample space is $\{0 \le \theta \le 2\pi\}$ and the function is $\theta = 1/2\pi$. In the case of a CRV, suppose we ask, What is the probability $\theta = 20°$. The spontaneous answer would be *zero*. But if we want $P(10° \le \theta \le 20°)$, then it is $1/36$. The idea of presenting a CRV this way is to illustrate the continuous nature of its **pdf**.

Consider another experiment of throwing a die or picking a card from a pack of 52. This is a DRV experiment for the simple reason that the sample space is finite and discrete. The **pdf** is discrete and exists only where an event occurs. If the earlier *roulette wheel* is mechanically redesigned so it will stop *only* at $\theta = \{0°, 1°, 2°, \ldots, 359°\}$, then θ becomes a DRV.

2.3.1 Functions of a Random Variable

Let us examine a few important relationships between functions of random variables. Most of the noise propagation in linear systems can be best understood by applying these *golden rules* partially or in full:

1. Consider a simple relation $\mathbf{z} = \mathbf{x} + \mathbf{y}$, then the pdf of $\mathbf{z}$ is obtained by convolving the **pdf** of $\mathbf{x}$ and $\mathbf{y}$, $f_{\mathbf{z}}(z) = f_{\mathbf{x}}(x) \otimes f_{\mathbf{y}}(y)$.
2. For the relation $\mathbf{y} = \mathbf{mx} + \mathbf{c}$, the **pdf** of $\mathbf{y}$ is $f_{\mathbf{y}}(y) = \frac{1}{m}f_{\mathbf{x}}\left(\frac{x-c}{m}\right)$.
3. The *central limit theorem* says that if $\mathbf{x}_i$ is a random variable of any distribution, then the random variable $\mathbf{y}$ defined as $\mathbf{y} = \sum_{i=1}^{N}(\mathbf{x}_i)$ has a *normal* distribution, where N is a large number. As an engineering approximation, $N \geq 12$ will suffice.

2.3.1.1 Hit Distance of a Projectile

This problem comes under non-linear functions of a random variable. To illustrate the idea, consider a ballistic object moving under the gravitational field shown in Figure 2.8 with a gravitational constant $g = 9.81 \, \mathrm{m/s^2}$:

$$v_x = v\cos\theta \quad \text{and} \quad v_y = v\sin\theta - gt, \tag{2.26}$$

$$x = vt\cos\theta \quad \text{and} \quad y = vt\sin\theta - \frac{1}{2}gt^2. \tag{2.27}$$

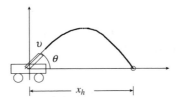

Figure 2.8   Ballistic motion

Setting $y = 0$ we get $t = (2v/g)\sin\theta$ for $t \neq 0$, then the horizontal distance where it hits the ground is

$$x_h = v\cos\theta \frac{2v\sin\theta}{g} = v^2\left(\frac{2\sin\theta\cos\theta}{g}\right) = \frac{v^2\sin 2\theta}{g}.$$

Consider v as statistically varying with $\mu_v = 50$ and $\sigma_v = 0.2$ and $\mu_\theta = \pi/3$ and $\sigma_\theta = \pi/30$ We have derived a closed-form solution (Figure 2.9) for the above pdf in the interval $85 < x < 250$:

$$f(x) = -\frac{7.61x^5}{10^{12}} + \frac{5.73x^4}{10^9} - \frac{1.66x^3}{10^6} + \frac{2.34x^2}{10^4} - 0.016x + 0.4259. \tag{2.28}$$

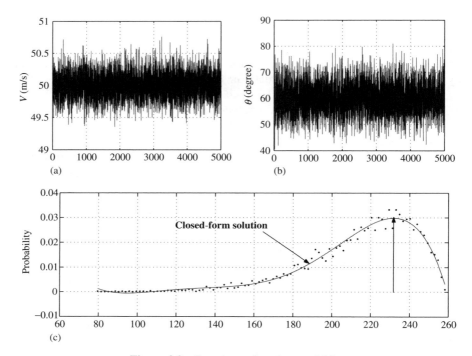

Figure 2.9 Functions of random variables

In this case we could find out the desired pdf but it is only because we have a closed-form solution for x_h as $(v^2/g)\sin 2\theta$.

2.3.2 Reliability of Systems

Systems are designed from components having different failure probabilities. A complex system is a function of many random variables and the study of these interconnected subsystems constitute an important branch of engineering. This must always be borne in mind by every product designer and practising engineer. Reliability can be achieved only at the product design stage, not at the end, hence product designers should pay attention to reliability.

In the case of software, system failures occur due to existence of many paths due to different input conditions and it behaves as a *finite-state automaton*. It is almost impossible to exhaust all the outcomes for all conditions. In fact, this is why many practices have been introduced in testing at each level. Reliability of the software improves with use whereas the reliability of a mechanical system degrades with use. Figure 2.10 shows the reliability graph of any generic system. Notice there are three phases in the graph: infantile mortality, random failures and aging. In the case of software there are only two phases exist; the aging phase doesn't exist.

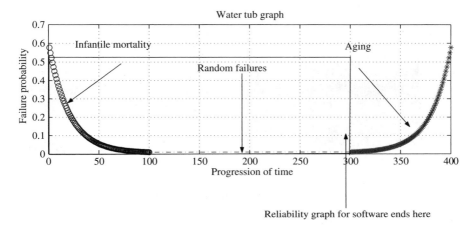

Figure 2.10   System reliability is a function of random variables

2.4 Noise

All textbooks of communication engineering and books on other disciplines define noise as an unwanted signal. Noise is characterised by two parameters: its amplitude statistics (probability density function) and its autocorrelation function r_k. Both of them are important and independent.

2.4.1 Noise Generation

On many occasions we need to generate specific types of noise. The most frequently used noise is white Gaussian noise (WGN). We cannot appreciate any noise in the time domain. The only way to get some information is by constructing the pdf of the given signal. There are many methods for generating WGN. The primary source is a uniformly distributed noise (UDN), which can be generated using a *linear congruential generator* (LCG), well explained by Knuth [10]. Figure 2.11 depicts the LCG relation given as $\mu_k = (a\mu_{k-1} + c) \bmod m$, where the constants a and c can be chosen to suit the purpose.

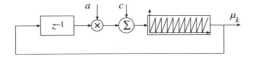

Figure 2.11   Linear congruential generator for UDF

WGN is generated by invoking the central limit theorem and also recollecting that the variance of UDN is $A^2/12$. Now WGN γ_k is given as $\gamma_k = \sum_{i=1}^{12} \mu_k$. Choosing 12 in the summation cancels the denominator 12 in the UDN μ_k.

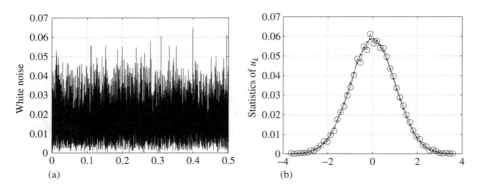

Figure 2.12   Amplitude statistics of u_k

2.4.2 *Fourier Transform and pdf of Noise*

We have constructed the pdf of a noise signal u_k that is white and Gaussian. It is given in Figure 2.12 for a WGN.

If the autocorrelation function $r_k = \delta_k$, then the noise is white. This is because the power spectrum of the noise is flat or the power level at all frequencies is constant. This whiteness (Figure 2.12(a)) is independent from the **pdf** of the noise (Figure 2.12(b)). The implication of this statement is that the **pdf** can be of any form. When the amplitude statistics of the given time series have a Gaussian distribution, and the power spectrum of the same signal has power levels constant and uniform across all the frequencies, then it is called white Gaussian noise, which essentially means that $r_k = \delta_k$.

2.5 Propagation of Noise in Linear Systems

When a WGN (Figure 2.12) is passed through a filter, the output is a coloured noise as in Figure 2.13(a). Here the amplitude statistics continue to remain Gaussian, as in Figure 2.13(b).

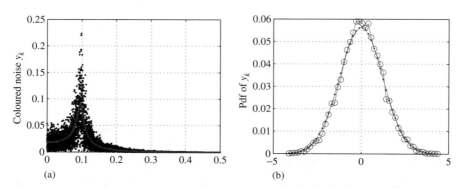

Figure 2.13   Amplitude statistics and power spectrum of y_k

Coloured Gaussian noise has identical structure or shape to the input (u_k) amplitude statistics, as in Figure 2.12(b). The output is said to be coloured, because spectral energy is not constant across all the frequencies, as shown in Figure 2.13(a)

Taking linear systems having either an IIR or an FIR and exciting them with Gaussian noise results in another Gaussian noise of different mean and variance. This is an important characteristic exhibited by linear systems and is illustrated in Figures 2.12 and 2.13. In this context, it is worth recollecting that a sine wave input to a linear system results in another sine wave with the same frequency but different amplitude and phase.

2.5.1 Linear System Driven by Arbitrary Noise

Consider an AR system defined by (2.23) and (2.24) and where the input u_k is uncorrelated and stationary. The output of this system is normal. As an example, let us excite the system with u_k having a Weibull distribution as in Figure 2.14 with $\alpha = 0.1$ and $\beta = 0.8$:

$$f(u) = \begin{cases} (\alpha/\beta^{\alpha})u^{\alpha-1}e^{-(u/\beta)^{\alpha}}, & u \geq 0, \\ 0, & u < 0. \end{cases} \tag{2.29}$$

The output is normal, as shown in Figure 2.14. This can be understood by realising that the output can be best approximated using an MA process as in (2.5) and (2.6) and that the statistics of u_k are the same as the statistics of u_{k+i}. Then use the three rules in Section 2.3.1 to obtain the output as a normal distribution.

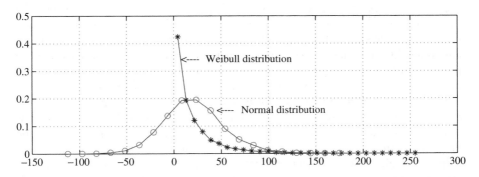

Figure 2.14   Input and output statistics of u_k and y_k

2.6 Multivariate Functions

Understanding functions of several variables is important for estimating parameters of systems. Often the problem is stated like this: Given a system, find out its response

to various inputs. This is a fairly simple problem known as *synthesis*. Given
the input and output, finding the system is more difficult; plenty of methods
are available and they depend on the specific nature of the problem. This section
tries to lay foundations for understanding parameters in more than one or two
dimensions.

It is the limitation of the mind that makes it difficult to perceive variables beyond
three dimensions. Even a school student can understand two dimensions with ease. A
point in the Cartesian frame is represented by an ordered pair with reference to x_1 and
x_2 axes, which are orthogonal to each other (Figure 2.15). Our mind is so well
trained it can understand and perceive all the concepts in two dimensions. The
difficulty comes once we go beyond two dimensions, and this puts the mind under
strain.

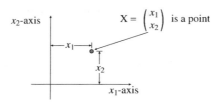

Figure 2.15   Two-dimensional representation

2.6.1 Vectors of More Than Two Dimensions

This section presents a novel way of visualising multidimensional variables. The
traditional approach is to imagine a hyperspace and treat it as a point. But we will
present a simple way of looking at the multidimensional vector, yet without losing
any generality. Consider a four-dimensional vector. $\mathbf{x}^t = [x_1, x_2, x_3, x_4]$ has been
represented as a polygon of 5 sides in Figure 2.16. This provides more insight into
understanding parameters of many dimensions. A multidimensional vector is a
figure but not a *point*. Visualising an *n*-dimensional vector as a *polygon* of order
$n + 1$ is an important aspect of understanding multivariate systems. This shows how
difficult it is to compare two figures or perform any standard binary operations on
them. Many times we need to find out maxima or minima in this domain. This will
help us to understand these concepts better. Comparing numbers or performing the
familiar binary operations such as $+, -, \div, \times$ is easy but it is more difficult with
multidimensional vectors as they are polygons. It doesn't mean that we cannot
compare two polygons; there are methods for doing this.

2.6.2 Functions of Several Variables

Multivariate functions are so common in reality, we cannot avoid handling them.
They could be representing a weather system or a digital filter; the motion
parameters of a space shuttle, a missile or an aircraft; the performance of a person

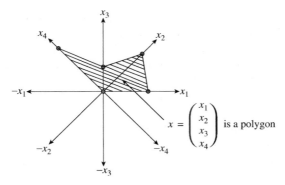

Figure 2.16   Representing a multidimensional vector

or company; the financial stability of a bank; and so on. The problem is to compare these multivariate functions by making judgements about their maxima or minima. In this context, understanding the n-dimensional vector gives a better solution to the problems. Control engineers and statisticians [7] have used these functions extensively.

Consider a multivariate function $y_k = f(\mathbf{x}_k, u_k)$ where $\mathbf{x}_k$ is a multidimensional vector and u is the input. This function describes a system. There are many times we need to drive the given system in a specific trajectory in a multidimensional space, subject to some conditions.

A digital filter defined in (2.23) and (2.24) is the best example of such functions and is written like this:

$$\mathbf{w}_k = \mathbf{A}\mathbf{w}_k z^{-1} + \mathbf{B}u_k \quad \text{and} \quad y_k = \mathbf{C}^t\mathbf{w}_k z^{-1} + \mathbf{D}u_k.$$

2.6.3 System of Equations

Another example of a multivariate function is a system of simultaneous equations. For the sake of brevity and comprehension, let us consider the following equations. Equations of this type are encountered in different ways in many applications. Pay attention to this subsection and it will give you a good insight into many solutions.

$$\begin{pmatrix} 0 & -5 \\ -8 & -6 \\ 3 & 2 \\ 0 & -9 \\ -5 & -3 \\ 4 & -4 \\ -8 & -9 \\ 9 & -8 \end{pmatrix} \begin{pmatrix} x_1 \\ x_2 \end{pmatrix} = \begin{pmatrix} -34 \\ -5 \\ 0 \\ -63 \\ 5 \\ -49 \\ -26 \\ -101 \end{pmatrix}, \tag{2.30}$$

$$\mathbf{Ax} = \mathbf{y}. \tag{2.31}$$

Equation (2.31) is a concise matrix way of representing the system of equations given in (2.30). In this particular case, matrix $\mathbf{A}$ is 8×2. The inverse of a matrix is defined only when $\mathbf{A}$ is square and $|\mathbf{A}| \neq 0$. A very innovative approach was adapted by Moore and Penrose for inverting a rectangular matrix.

Take (2.31) and premultiply by $\mathbf{A}^t$ to obtain

$$\mathbf{A}^t\mathbf{A}\mathbf{x} = \mathbf{A}^t\mathbf{y}. \tag{2.32}$$

The matrix $\mathbf{A}^t\mathbf{A}$ is square and can also be written as[4]

$$\mathbf{A}^t\mathbf{A} = \begin{pmatrix} 0 \\ -5 \end{pmatrix}(0 \quad -5) + \begin{pmatrix} -8 \\ -6 \end{pmatrix}(-8 \quad -6) + \cdots + (9-8)(9 \quad -8)$$

$$= \sum_{i=1}^{8} \mathbf{x}_i\mathbf{x}^t \quad (\mathbf{x}_i \text{ is } i\text{th row vector of } \mathbf{A}). \tag{2.33}$$

The column vector $\mathbf{A}^t\mathbf{y}$ can be written as

$$\mathbf{A}^t\mathbf{y} = -34\begin{pmatrix} 0 \\ -5 \end{pmatrix} - 5\begin{pmatrix} -8 \\ -6 \end{pmatrix} + \cdots - 101\begin{pmatrix} 9 \\ -8 \end{pmatrix}$$

$$= \sum_{i=1}^{8} \mathbf{x}_i^t y_i. \tag{2.34}$$

Under the assumption that $\mathbf{A}^t\mathbf{A}$ is non-singular, the inverse can be obtained recursively or en block. Then the solution for the system (2.30) is given as

$$\mathbf{x} = [\mathbf{A}^t\mathbf{A}]^{-1}\mathbf{A}^t\mathbf{y} = \mathbf{A}^+\mathbf{y}, \tag{2.35}$$

where $\mathbf{A}^+ \triangleq [\mathbf{A}^t\mathbf{A}]^{-1}\mathbf{A}^t$ is known as the Moore–Penrose pseudo-inverse [1, 8]. The solution is $\mathbf{x} = \mathbf{A}^+\mathbf{y}$. There are two ways of looking at it. One way is that it corresponds to the best intersection [9] of all the intersections in Figure 2.17, in a

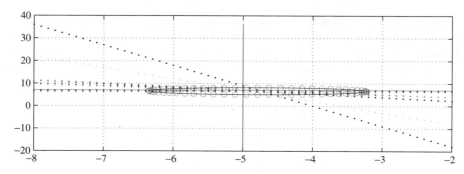

Figure 2.17 A system of linear equations

---

[4]This is useful in inverting recursively using a matrix inversion lemma which states that $(\mathbf{B} + \mathbf{x}\mathbf{x}^t)^{-1} = \mathbf{B}^{-1} - \left(\frac{\mathbf{B}^{-1}\mathbf{x}\mathbf{x}^t\mathbf{B}^{-1}}{1+\mathbf{x}^t\mathbf{B}^{-1}\mathbf{x}}\right)$.

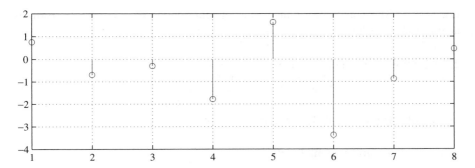

Figure 2.18 Innovations e_k

least-squares sense. The other way is that the rows of $\mathbf{A}^+$ corresponds to MA processes or FIR filters. Figure 2.18 depicts the error e_k (in simple language it is the deviation from the straight line fit). Sometimes ϵ_k is called an innovation process [4] and is given as $e_k = \mathbf{A}\mathbf{A}^+\mathbf{y} - \mathbf{y}$ (Figure 2.18).

2.7 Number Systems

A good understanding of binary number systems provide a proper insight into digital filters and their implementations. Consider a binary number $b_n = x_n x_{n-1} \ldots x_3 x_2 x_1$, where the symbol x_i, called a *bit* of the *word* b_n, takes value 0 or 1. The string of *binary* symbols has a value given as

$$b_n = x_n x_{n-1} \ldots x_i \ldots x_3 x_2 x_1 = \sum_{i=1}^{n} x_i 2^{i-1}. \qquad (2.36)$$

This representation is an unsigned representation of integer numbers. For a given value of n we can generate $2^n - 1$ patterns. If we fix the value n, normally known as word length, to accommodate the positive and negative numbers, the usable numbers gets partitioned into half each. There are three common ways of doing it. For illustration we choose the value $n = 3$.

2.7.1 Representation of Numbers

2.7.1.1 Sign Magnitude Form

Let us consider a set of 8 numbers $\{0, 1, 2, 3, 4, 5, 6, 7\}$ representing a 3-bit binary numbers and mapped as $\{+0, +1, +2, +3, -0, -1, -2, -3\}$. Here we have divided the given set into two partitions of positive numbers and negative numbers. Human beings understand decimal numbers the best, so we are trying to explain with equivalent decimal numbers whose binary bit patterns are the same. The basic problem is that the hardware is relatively complex and the non-uniqueness of $+0$

and -0 is not acceptable in the mathematical sense. Sign magnitude form is abbreviated to SM.

2.7.1.2 Ones Complement Form

In ones complement form (1sC) the same numbers $\{0, 1, 2, 3, 4, 5, 6, 7\}$ are designated as $\{+0, +1, +2, +3, -3, -2, -1, -0\}$. This method is relatively easy to implement but still has the problem of a non-unique zero.

2.7.1.3 Twos Complement Form

Twos complement form (2sC) is the most popular numbering and is practised widely in all machines. In this scheme the positive zero and negative zero are removed and the numbers $\{0, 1, 2, 3, 4, 5, 6, 7\}$ are mapped as $\{0, 1, 2, 3, -4, -3, -2, -1\}$.

2.7.1.4 The Three Forms in a Nutshell

Table 2.1 illustrates the three representations in a nutshell. Sign magnitude is very popular in A/D converters and it is almost a standard practice in all arithmetics to use a twos complement number system.

Table 2.1   The three representations in a nutshell

$\lfloor x_k \rfloor$	SM	1sC	2sC
000	$+0$	$+0$	0
001	$+1$	$+1$	1
010	$+2$	$+2$	2
011	$+3$	$+3$	3
100	-0	-3	-4
101	-1	-2	-3
110	-2	-1	-2
111	-3	-0	-1

2.7.2 Fixed-Point Numbers

Similar to the imaginary decimal point, we have a *binary* point that can be positioned anywhere in the binary number b_n (2.36) and can be written as

$$b_n = \boxed{x_n} \cdot \boxed{x_{n-1} \ldots x_i \ldots x_3 x_2 x_1}.$$

Now the value of b_n takes the form of a rational number $b_n/2^{n-1}$; all other things remain the same. It could be **SM, 1sC** or **2sC**. The above form is called the *normalised* form. It essentially imposes a restriction that it is permitted to have only one *bit* to the left of the imaginary binary point

2.7.3 Floating-Point Numbers

Fixed-point representation suffers from the limitation of *dynamic range*, hence the floating point has evolved to give

$$b_n = x_n m \times 2^c,$$

where m is the *mantissa* and c is the *characteristic* or *exponent*; x_n is the sign *bit*. The floating-point number is represented as a 3-tuple given as $b_n = \{x_n, m, c\}$ in packed form. The standard[5] IEEE format for 32-bit single-precision floating-point number is given as

$$b_n = \underset{x_n}{\langle x_{32} \rangle} \; \underset{c}{\langle x_{31} \dots x_{24} \rangle} \; \underset{m}{\langle x_{23} \dots x_1 \rangle}$$

In this notation, m is a fixed-point positive fractional number in *normalised* form and the value b_n represented by the word may be determined as follows:

1. If $0 < c < 255$ then $b_n = -1^{x_n} \times 2^{c-127} \times (1.m)$ where $1.m$ is intended to represent the binary number created by prefixing m with an implicit leading 1 and a binary point.
2. If $c = 0$ and $m \neq 0$, then $b_n = -1^{x_n} \times 2^{-126} \times 0.m$ These are *unnormalised* values.
3. If $c = 255$ and $m \neq 0$, then $b_n = $ NaN (Nan means not a number).
4. If $c = 255$ and $m = 0$ and $x_n = 1$ or 0, then $b_n = -\infty$ or $b_n = \infty$.
5. If $c = 0$ and $m = 0$ and $x_n = 1$ or 0, then $b_n = -0$ or $+0$

2.8 Summary

This chapter was a refresher on some relevant engineering topics. It should help you with the rest of the book. We considered the autocorrelation function, representation of linear systems, and noise and its propagation in linear systems. We also discussed the need to know about systems reliability. Most problems in real life are inverse problems, so we introduced the Moore–Penrose pseudo-inverse. We concluded by providing an insight into binary number systems.

References

1. S. Slott and L. James, *Parametric Estimation as an Optimisation Problem*. Hatfield Polytechnic.
2. D. G. Luenberger, *Introduction to Linear and Non-linear Programming*. Reading MA: Addison-Wesley, 1973.
3. T. L. Boullion and P. L. Odell, *Generalised Inverse Matrices*. New York: John Wiley & Sons, Inc., 1971.

---

[5] ANSI/IEEE Standard 754-1985, Standard for Binary Floating Point Arithmetic.

4. A. V. Openheim, A. S. Wilsky and I. T. Young, *Signals and Systems*. Englewood Cliffs NJ: Prentice Hall, 1983.
5. R. Isermann, *Digital Control Systems*. Berlin: Springer-Verlag, 1981.
6. J. R. Wolberg, *Prediction Analysis*. New York: Van Nostrand, 1967.
7. S. Brandt, *Statistical and Computational Methods in Data Analysis*. Dordrecht: North Holland, 1978.
8. A. Albert, *Regression and the Moore–Penrose Pseudoinverse*. New York: Academic Press, 1972.
9. P. Lancaster, The Theory of matrices: With Applications, 2nd edn.
10. Donald E. Knuth, *Art of Computer Programming*.

3

Digital Filters

We use digital filters [1] on many occasions without our knowledge. The main reason for this is that any sequence of numbers if multiplied[1] and added constitutes a simple linear filter. The temperature of a bulb follows a similar law but with some more modifications. Computing compound interest has a similar structure.

Such filters are nothing but a difference equation with or without feedback. For a software engineer it is a piece of code; for a hardware engineer it is an assembly of shift registers, adders, multipliers and control logic.

Converting a difference equation to a program is fairly simple and routine. The key step is to choose the coefficients of the difference equation so they suit your purpose. There are many methods available in the published literature [1, 2] for obtaining the coefficients in these difference equations. There are also many powerful software packages[2] available to assist the designer in the implementation.

There is another class of design approach required in many practical situations, where we need to obtain the filter coefficients given the measured impulse response of the filter (h_k), or given an input (u_k) and an output sequence (y_k), or given the amplitude and phase response of the filter. Problems of this type are known as *inverse* problems. For a difference equation, obtaining the output corresponding to a given input is relatively easy and is known as the *forward* problem.

3.1 How to Specify a Filter

Conventionally filters are specified in the frequency domain. For a better understanding, consider the filter in Figure 3.1. There are three important parameters that often need to be specified when designing a filter. Conventionally, slope or roll-off is defined in dB/decade or dB/octave.

---

[1]By another set of constants.
[2]MATLAB, SABER, etc.

Digital Signal Processing: A Practitioner's Approach   K. V. Rangarao and R. K. Mallik
© 2005 John Wiley & Sons, Ltd

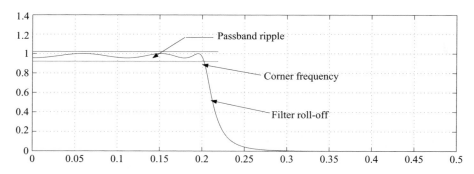

Figure 3.1 Filter specifications

The unit dB/octave is from music and acoustics. The unit dB/decade is from control engineering; it is slowly becoming more prevalent in filter design. Most of the programs available demand the order of the filter. A simple rule of thumb is if N is the order of the system, then the corresponding slope at the cut-off frequency is $20N$ dB/decade. A quick reference to Figure 3.1 shows a typical filter specification in the frequency domain. We specify the filter using

- Passband or stopband ripple in dB
- Corner frequency or cut-off frequency
- Roll-off defined in dB/decade

3.2 Moving-Average Filters

Moving-average (MA) filters are the simplest filters and can be well understood without much difficulty. The basic reason is that there is no feedback in these filters. The advantage is that they are always stable and easy to implement. On the other hand, filters with feedback elements but no feedforward elements are called autoregressive (AR) filters. Here are some examples.

3.2.1 Area under a Curve

The most common MA filters are used for finding the area under a portion of a curve by the trapezium rule or Simpson's rule, represented as

$$\bar{u}_k = \frac{u_k + u_{k-1}}{2} \qquad \text{trapezium rule,} \qquad (3.1)$$

$$\bar{u}_k = \frac{u_k + 4u_{k-1} + u_{k-2}}{3} \qquad \text{Simpson's rule,} \qquad (3.2)$$

where $\{u_k\}$ is the input and $\{\bar{u}_k\}$ the output. The objective of these two filters is to find the area between the points by finding out the weighted mean ($\bar{u}_k$) of the ordinates.

3.2.2 Mean of a Given Sequence

Finding the mean of a given sequence is another application for an MA filter. However, the mean of a sequence of length N can also be found recursively, resulting in an AR filter. We derived this in (1.41). It is given as

$$\mu_k = \left(\frac{1}{1 + 1/N}\right)\mu_{k-1} + \left(\frac{1}{N+1}\right)u_k \tag{3.3}$$

$$= a_1\mu_{k-1} + b_1 u_k \quad \text{where} \quad a_1 + b_1 = 1. \tag{3.4}$$

Another filter commonly used is an exponential weighted-averaging filter with similar structure (also known as a *fading* filter); it is given as

$$y_k = \lambda y_{k-1} + (1 - \lambda)u_k. \tag{3.5}$$

This filter is used for finding out the mean when distant values are of no significance. In fact there is no difference between the filters in (3.4) and (3.5); it is only the contexts that are different. The solution for (3.5) is of the form $y_k = (A\lambda^k) \otimes u_k$, where A is a constant that can be obtained from the initial conditions and $\otimes$ represents convolution. The general choice of the coefficient is $0.8 \leq \lambda \leq 0.95$.

3.2.3 Mean over a Fixed Number of Samples

The filter in (3.4) finds the mean of a given set of samples. However, on many occasions, we need to find continuously the mean over a fixed number of samples. This can be done using the filter in (3.5), but then the far samples are weighted low compared to the near samples.

Consider the case when we need to compute the mean (μ_k) over $n = k$ to $n = k + N - 1$ for the given time series u_k as shown in Figure 3.2. This is

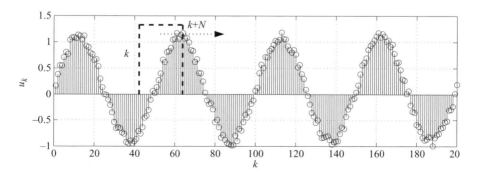

Figure 3.2 Rectangular moving filter

mathematically given as

$$\mu_k = \frac{1}{N} \left(\sum_{i=k}^{i=k+N-1} u_i \right). \tag{3.6}$$

To arrive at the recursive form similar to (3.4), we rewrite μ_{k+1} as

$$\begin{aligned}
N\mu_{k+1} &= \left(\sum_{i=k+1}^{i=k+N} u_i \right) + u_k - u_k \\
&= \left(\sum_{i=k}^{i=k+N} u_i \right) - u_k \\
&= \left(\sum_{i=k}^{i=k+N-1} u_i \right) + u_{k+N} - u_k \\
&= N\mu_k + u_{k+N} - u_k. \tag{3.7}
\end{aligned}$$

Equation (3.7) can be written as

$$\mu_k = \mu_{k-1} + ge_k \quad \text{where} \quad g = \frac{1}{N} \quad \text{and} \quad e_k = u_k - u_{k-N}. \tag{3.8}$$

The moving mean μ_k is obtained by passing the signal ge_k through an *integrator* (3.8). This integrator can be written in operator notation or as a transfer function (TF) model:

$$\mu_k = u_k \left[\frac{1}{N} \left(\frac{1 - z^{-N}}{1 - z^{-1}} \right) \right]. \tag{3.9}$$

The filter in (3.9) has feedback and feedforward elements and is called an autoregressive moving average (ARMA) filter. This filter is *stable* in spite of a pole on the unit circle due to the pole–zero cancellation at $z = 1$.

3.2.4 Linear Phase Filters

Consider a narrowband filter (NBF) given by (3.17). Let the impulse response of this IIR filter be h_k. We truncate the infinite series h_k and use these values to generate the coefficients of an MA filter in a specific way, given by

$$\begin{aligned}
b_i &= h_{N-i} \quad \text{for} \quad i = 1 \text{ to } N, \\
b_{N+i} &= h_{i+1} \quad \text{for} \quad i = 1 \text{ to } N - 1. \tag{3.10}
\end{aligned}$$

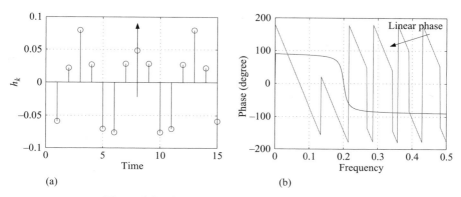

Figure 3.3   Coefficients of a moving-average filter

The coefficient vector **b**, comprising the b_i stacked together, is of dimension $2N - 1$; it is shown in Figure 3.3(a). Notice that the coefficients are symmetric; we have drawn an arrow at the symmetric point.

From the theory of polynomials, we recollect that such a symmetric property produces a linear phase characteristic. The same is shown in Figure 3.3(b), which shows the linear phase of the MA filter. In the same figure, we see the corresponding non-linear phase of the original (parent) IIR narrowband filter. The choice of these coefficients is deliberate and is intended to describe the linear phase property. In addition, we demonstrate conversion of an IIR filter to an MA filter using the *truncated* impulse response of the IIR filter.

Figure 3.4 shows the frequency response of the IIR filter and the MA filter (truncated IIR). The performance of the IIR filter is far superior to the performance of the MA filter. Computationally, an IIR filter needs less resources, but stability and phase are the compromises. In addition, a careful examination reveals a compromise in the filter's *skirt* sharpness and its sidelobe levels are high.

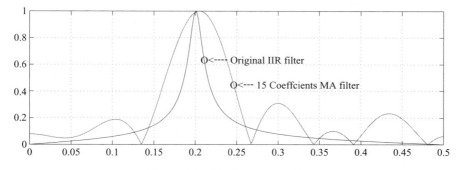

Figure 3.4   Moving-average filter

3.2.5 MA Filter with Complex Coefficients

There is another important breed of MA filters based on the concept of super-heterodyning. This consists of a local oscillator, a mixer, and a filter. The signal $x(t)$ is multiplied with a local oscillator $e^{-j2\pi ft}$ and passed through an integral or lowpass filter. Mathematically,[3] it is given by

$$\frac{U}{a_k + jb_k} = \frac{\int u(t)\,e^{-j2\pi ft}\,dt}{\sum u_k\,e^{-j2\pi fk}} \tag{3.11}$$

Equation (3.11) depicts a mapping from continuous to discrete. This means that we *beat* the given signal u_k with an oscillator and pass the result through a summation. This results in a pair of MA filters:

$$a_k = \sum u_k \cos(2\pi f\,k) \quad \text{and} \quad b_k = \sum u_k \sin(2\pi f\,k). \tag{3.12}$$

Figure 3.5 is an implementation of (3.12) with non-linear devices such as multipliers.

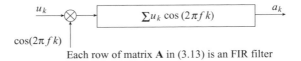

Each row of matrix **A** in (3.13) is an FIR filter

Figure 3.5   Heterodyning

When we sweep the oscillator to generate a set of rows resulting in a pair of matrices, one for real (in-phase or cosine) and one for imaginary (quadrature or sine), we can also observe the output as Fourier series coefficients.

Instead of using an infinite sine wave, we can sample the sine wave to generate the coefficients of the MA filter. The filter characteristics depend on the number of coefficients and how many samples are taken per cycle. Consider a matrix **A** with each row containing the coefficients of an MA filter.

$$\mathbf{A} = \begin{pmatrix} 1 & 1 & 1 & 1 & 1 & 1 & 1 & 1 \\ 1 & 0.7071 & 0 & -0.7071 & -1 & -0.7071 & 0 & 0.7071 \\ 1 & 0 & -1 & 0 & 1 & 0 & -1 & 0 \\ 1 & -0.7071 & 0 & 0.7071 & -1 & 0.7071 & 0 & -0.7071 \\ 1 & -1 & 1 & -1 & 1 & -1 & 1 & -1 \\ 1 & -0.7071 & 0 & 0.7071 & -1 & 0.7071 & 0 & -0.7071 \\ 1 & 0 & -1 & 0 & 1 & 0 & -1 & 0 \\ 1 & 0.7071 & 0 & -0.7071 & -1 & -0.7071 & 0 & 0.7071 \end{pmatrix} \tag{3.13}$$

---

[3]Note that we have avoided using limits.

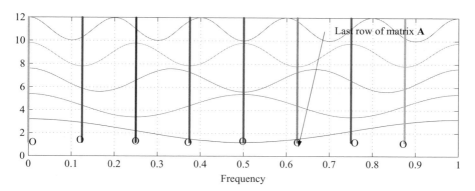

Figure 3.6 How to understand matrix **A** in (3.12)

We get a better insight into the entries of the matrix **A** by looking at Figure 3.6. Look back and forth between Figures 3.6 and 3.5 and glance at the matrix (3.13). This will help you understand what is going on. A close look at Figure 3.6 shows that cosine waves are sampled at 8 places simultaneously, leading to various rows in the matrix **A**.

Note that each row represents an oscillator. The choice of the coefficients is not arbitrary but has a definite pattern. The last row of **A** consists of 8 samples of one full cycle of a cosine wave. We have two matrices, one for the real part and one for the complex part. The entries of each row are sampled cosine values or sine values. The given input is passed through these two filters and this creates two orthogonal outputs. It is not necessary to have a square matrix, but there is a computational advantage in choosing a square matrix of order 2^n.

All these filters have linear phase characteristics. Consider the row vector $\mathbf{a}_i$, the ith row of the matrix **A**. The output $y_k^i = \mathbf{a}_i u_k$ is an MA process.

$$
\mathbf{B} = \begin{pmatrix}
0 & 0 & 0 & 0 & 0 & 0 & 0 & 0 \\
0 & -0.7071 & -1 & -0.7071 & 0 & 0.7071 & 1 & 0.7071 \\
0 & -1 & 0 & 1 & 0 & -1 & 0 & 1 \\
0 & -0.7071 & 1 & -0.7071 & 0 & 0.7071 & -1 & 0.7071 \\
0 & 0 & 0 & 0 & 0 & 0 & 0 & 0 \\
0 & 0.7071 & -1 & 0.7071 & 0 & -0.7071 & 1 & -0.7071 \\
0 & 1 & 0 & -1 & 0 & 1 & 0 & -1 \\
0 & 0.7071 & 1 & 0.7071 & 0 & -0.7071 & -1 & -0.7071
\end{pmatrix}
\tag{3.14}
$$

The matrix **B** in (3.14) corresponds to the imaginary part of the coefficients; the rest of the explanation is the same as for **A**. Figure 3.7 shows the in-phase and quadrature results for matrices **A** and **B**.

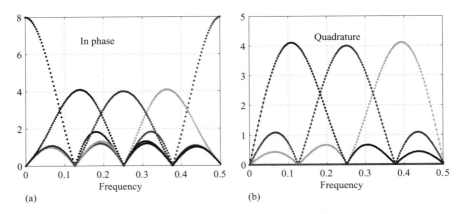

Figure 3.7   Bank of filters in matrices **A** and **B** in (3.13) and (3.14)

3.3 Infinite Sequence Generation

In many applications, we need to generate an infinite sequence of numbers for testing or as part of the system modelling. Here are a few examples.

3.3.1 Digital Counter

In many programing applications we use counters. Such counters are obtained by using an *integrator* with an input as a Kronecker delta function given by $\delta_k = 1$ for only $k = 0$, and the input–output relation

$$y_k = y_{k-1} + \delta_k. \tag{3.15}$$

Quite often this model is used in real-time simulations for generating a ramp.

3.3.2 Noise Sequence

The primary source is a uniformly distributed noise (UDN), which can be generated by a *linear congruential generator* (LCG), explained in Section 2.4.1. Figure 3.8

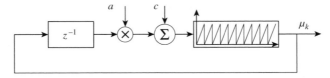

Figure 3.8   Non-linear infinite sequence generator

shows the relation of an LCG given as $\mu_k = (a\mu_{k-1} + c) \bmod m$, where the constants a and c can be chosen to suit the purpose. This is an example of a filter with a non-linear element in the feedback loop.

3.3.3 Numerically Controlled Oscillator

Consider this statement:

> For a sequence of numbers, multiply the previous number by 1.8 and subtract the previous-previous number to result in a new number. Start with a previous number of 1 and a previous-previous number of 0.

Translated into an equation, this is

$$y_k = 1.8y_{k-1} - y_{k-2} \quad \text{given} \quad y_0 = 1 \quad y_{-1} = 0. \tag{3.16}$$

Equation (3.16) results in the infinite sequence of Figure 3.9; it is a sinusoidal wave with a normalised frequency $f_n = \frac{1}{2\pi}\cos^{-1}\frac{1.8}{2} = 0.0718$. The infinite

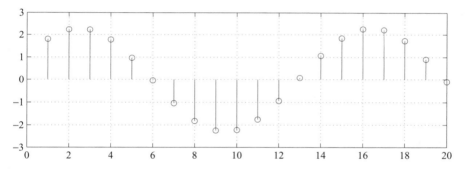

Figure 3.9 Numerically controlled oscillator

sequence $\{y_k\}$ generated by the 3-tuple $\{1.8, 1, 0\}$ illustrates a method for data compression. In addition, by varying only the coefficient of y_{k-1}, we can control the frequency; this is an important result used in the implementation of a numerically controlled oscillator (NCO). Consider a simple pendulum oscillating. If we watch this using a stroboscope, the angle seen at the time of flash is $\theta_k = \kappa\theta_{k-1} - \theta_{k-2}$, where κ is a constant that depends on the flash frequency, the length of the pendulum, and the acceleration due to gravity at that place.

3.4 Unity-Gain Narrowband Filter

One of the most popular filters is the narrowband filter (NBF). It is used in every domain of filter design. All filter designs of higher order use cascades of NBFs having different parameters. In superheterodyning receivers, the NBF is used immediately after the mixer stage. It is also very common in notch filters. In such filters there is a need to have unity gain around the peak response and zero gains at DC and at the Nyquist frequency, resulting in an input–output relation

$$y_k = rpy_{k-1} - r^2 y_{k-2} + g(u_k - u_{k-2}), \qquad (3.17)$$

where $p = 2 \cos \theta$, with $\theta = 2\pi f$ and $g = (1 - r^2)/2$. We note that $-2 \leq p \leq 2$. For stability, we need $0 < r < 1$. The frequency response of the NBF is given in Figure 3.10. It has gain zero at 0 and 0.5 due to the two zeros present at these places, with $\mathbf{a} = [1 - rp\, r^2]$ and $\mathbf{b} = g[1\,0 - 1]$. Equation (3.17), for an NBF, becomes an NCO equation when we substitute $r = 1$.

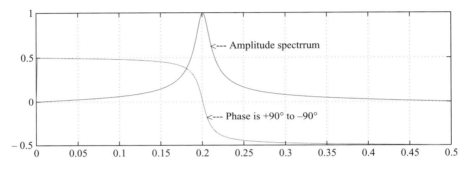

Figure 3.10   Bandpass filter with $r = 0.95$ and $f = 0.2$

3.5 All-Pass Filter

Many times we need to control the phase of the sequence or need to interpolate the values of a given sequence. This is best done by filters of first order or second order. In the first-order filter, we need to remember just *one* old value; in the second-order filter, we need to remember *two* old values. A typical first-order all-pass filter (APF) has the structure

$$y_k = ry_{k-1} + ru_k - u_{k-1}. \qquad (3.18)$$

The second-order filter is

$$y_k = rpy_{k-1} - r^2 y_{k-2} + r^2 u_k - rpu_{k-1} + u_{k-2}. \qquad (3.19)$$

These two all-pass filters represent a real-pole filter and a complex-pole filter, respectively. When we have complex poles, the roots bear a reciprocal conjugate relation. This can be best understood by writing (3.19) in operator notation:

$$(1 - re^{j\theta}z^{-1})(1 - re^{-j\theta}z^{-1})y_k = \left(1 - \frac{1}{r}e^{j\theta}z^{-1}\right)\left(1 - \frac{1}{r}e^{-j\theta}z^{-1}\right)u_k. \qquad (3.20)$$

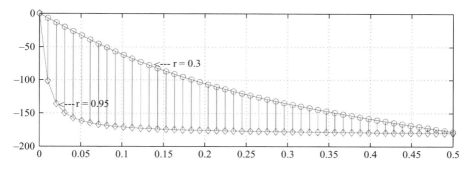

Figure 3.11 Phase of a first-order all-pass filter

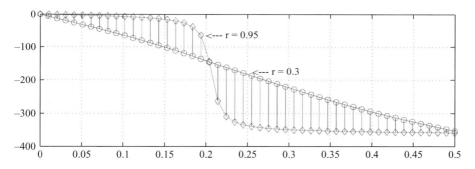

Figure 3.12 Phase of a second-order all-pass filter

The phase shown in Figure 3.11 is for a real-pole filter. Notice the sensitivity of r in controlling phase, and notice that between 50 samples/cycle and 4 samples/cycle the filter is very effective at controlling the phase. Figure 3.12 shows the phase of a complex all-pass filter and its sensitivity to the radial pole position.

3.5.1 Interpolation of Data Using an APF

A typical data acquisition system with more than one signal has the structure given in Figure 3.13. This structure is chosen for economy and to make best use of the A/D converter's bandwidth.

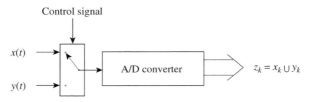

Figure 3.13 Phase shift in the data acquisition

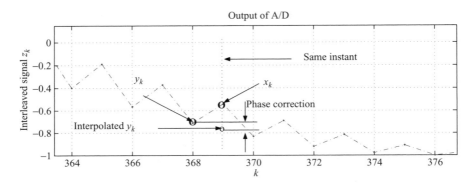

Figure 3.14 Correcting the phase shift

The signals $x(t)$ or[4] $y(t)$ are passed through an analogue multiplexer, with the help of a control signal. The digitised signals x_k and y_k are interleaved to yield $z_k = x_k \cup y_k$ and passed to the processor for further processing in the digital domain. Ideally, the requirement is to sample the signals at the same time. However, to reduce the amount of hardware, we decide to use only a single A/D converter. Refer to Figure 3.14; what we need is the interpolated value of $y_k = \hat{y}_k$ corresponding to x_k. It implies that we *should not* take the two sequences $\{x_k\}$ and $\{y_k\}$ for further processing, but the sequences $\{x_k\}$ and $\{\hat{y}_k\}$. And this implies three things:

1. The interleaved sequence z_k needs to be separated as x_k and y_k.
2. We reconstruct the sequence $\hat{y}_k$ from y_k by passing through an APF.
3. We use this pair of sequences $\{x_k\}$ and $\{\hat{y}_k\}$ as if they are sampled simultaneously.

Obtaining the sequence $\{\hat{y}_k\}$ is known as phase correction and is shown in Figure 3.14. This should be done for the entire sequence with linear phase characteristics. It is typically done using an APF.

3.5.2 Delay Estimation Using an APF

We demonstrate an adaptive APF for estimating the delay between two sinusoidal signals. An APF has a flat frequency response and a variable phase response. We best use this principle by varying the pole and zero positions. Consider two omnidirectional antenna elements spaced at a distance $d \leq \lambda/2$ to avoid the grating lobe phenomenon (similar to the aliasing problem in the time domain). Estimating the value of φ is important to find out the direction of an emitter producing narrowband signals.

---

[4]Mutually exclusive *or*.

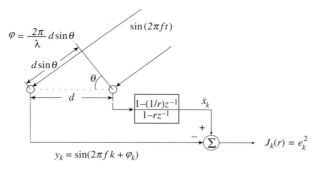

Figure 3.15 Delay estimation using an all-pass filter

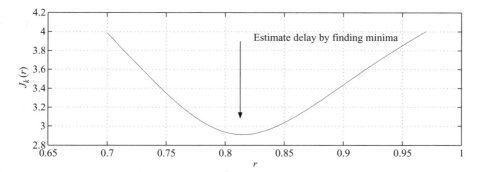

Figure 3.16 Varying the parameter r for delay estimation

The signal received at the first element after digitisation is $x_k = \sin(2\pi fk)$ and the signal at the second element is $y_k = \sin(2\pi fk + \varphi_k)$. The phase shift or delay $\varphi_k = (2\pi/\lambda)d\sin\theta$. If the speed of propagation is c, then $c = f\lambda$. We pass x_k through an APF which acts as interpolator or a variable delay filter with constant gain. We generate a criteria function $J_k(r) = \log(e_k^2)$, where $e_k = \tilde{x}_k - y_k$ and $\tilde{x}_k$ is the output of the APF. This is shown in Figure 3.15. Notice that the function in Figure 3.16 has a distinct minimum, so we can adapt the filter by giving a proper adaptation rule for delay estimation.

3.6 Notch Filter

Quite often we need to specifically eliminate some frequencies in the spectrum of the signal. This demands a notch filter [3, 4]. These applications include removal of unwanted signals in a specific band for proper operation of receivers in communication systems. This is realised using a simple model for the notch filter (NF) transfer function given by $NF = 1 - NBF$, where NBF denotes the unity-gain narrowband filter (NBF), i.e. $NF = 1 - B(z)/A(z)$. This leads to a simple difference form

$$y_k = rpy_{k-1} - r^2 y_{k-2} + \frac{1+r^2}{2} u_k - rpu_{k-1} + \frac{1+r^2}{2} u_{k-2}. \qquad (3.21)$$

The above notch filter can be written in a different way to understand it better by defining $\theta = 2\pi f$ and $\varphi = \cos^{-1}[rp/(1+r^2)]$ and rewriting the filter as

$$(1 - re^{j\theta}z^{-1})(1 - re^{-j\theta}z^{-1})y_k = (1 - e^{j\varphi}z^{-1})(1 - e^{-j\varphi}z^{-1})u_k. \qquad (3.22)$$

The physical significance is that θ is the natural resonant angular frequency, while at angular frequency φ the frequency response of the system is maximum for the second-order system. This helps to give a feel for its operation while implementing the filter.

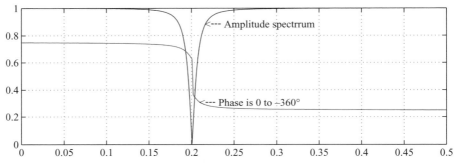

Figure 3.17   Notch filter

3.6.1 Overview

We have covered a variety of filters and we have seen that only the denominator polynomial makes a significant difference in the NBF, APF and NF; and even in the NCO, taking $r = 1$ is a point to be noted. A thorough understanding of first- and second-order systems is the basis of understanding any digital filter.

3.7 Other Autoregressive Filters

Autoregressive filters are very commonly used, and excellent software packages can obtain the coefficients for given filter specifications. Most of the popular filters, such as Chebyshev and Butterworth (Figure 3.18), are cascaded second-order filters of

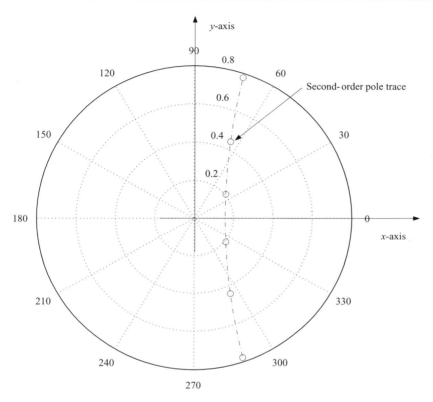

Figure 3.18   A Butterworth autoregressive filter

the type given in (3.17) with minor variations to meet the overall specifications. Consider a filter with the following specifications (the same as in Figure 3.1):

1. Cut-off frequency as $0.2\,\text{Hz}$;
2. Sixth-order filter giving a roll-off of $6 \times 20 = 120$ dB/decade
3. A permitted passband ripple of 0.4 dB

With these specifications, using the MATLAB package, we get the Chebyshev filter as

$$
\begin{aligned}
x_k = {} & 2.5579x_{k-1} - 3.9518x_{k-2} + 3.8694x_{k-3} - 2.5378x_{k-4} \\
& + 1.0415x_{k-5} - 0.2144x_{k-6} + 0.0035u_k + 0.021u_{k-1} \\
& + 0.0526u_{k-2} + 0.0702u_{k-3} + 0.0526u_{k-4} + 0.0210u_{k-5} \\
& + 0.0035u_{k-6}
\end{aligned}
\tag{3.23}
$$

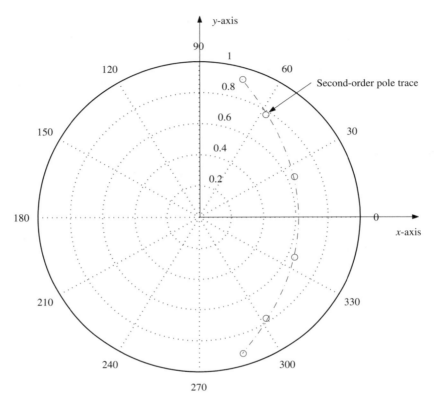

Figure 3.19 A Chebyshev autoregressive filter

A pole–zero plot of the same filter is shown in Figure 3.19. We can see that the desired filter is a cascade of second-order filters with poles following an equation $x - 0.6168 = -0.4478y^2$ and symmetric about the x-axis while three pairs of second-order zeros are positioned on the unit circle at 180^0. It is very interesting to note that the poles are placed on a *parabolic* path with focal point towards the origin.

A filter with the same specifications except a flat passband characteristic is obtained using the Butterworth model, which results in

$$
\begin{aligned}
x_k = {} & 0.1876x_{k-1} - 1.3052x_{k-2} + 0.6743x_{k-3} - 0.2635x_{k-4} \\
& + 0.0518x_{k-5} - 0.0050x_{k-6} + 0.0103u_k + 0.0619u_{k-1} \\
& + 0.1547u_{k-2} + 0.2063u_{k-3} + 0.1547u_{k-4} + 0.0619u_{k-5} \\
& + 0.0103u_{k-6}.
\end{aligned}
\tag{3.24}
$$

Unlike the pole–zero plot of the Chebyshev filter, we see that the focal point of the Butterworth filter plot is away from the origin. The pole trace follows the equation $x - 0.1582 = 0.1651y^2$, as shown in Figure 3.18. Once again, the zeros are positioned at $180°$ on the unit circle.

By now you have probably understood why we have focused on second-order filters. Almost all the filters can be realised by cascading second-order filters. For detailed design methods, consult a book on filter design.

3.8 Adaptive Filters

An Adaptive filter is a filter whose characteristics can be modified to achieve an objective by automatic adaptation or modification of the filter parameters. We illustrate this using a simple unity-gain BPF filter described in Section 3.4 via (3.17). We demonstrate this idea of adaptation by taking a specific criterion. In general, the criteria and the choice of feedback for adjusting parameters make an adaptive filter robust.

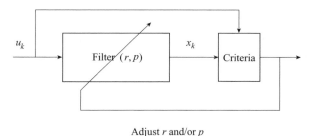

Adjust r and/or p

Figure 3.20   Adaptive filter

In this particular case, $-2 \le p \le 2$ since we have $p = 2\cos\theta$. The parameter r is in $(0,1)$. By choosing $r = 1$ the filter becomes marginally stable and this must be avoided.

3.8.1 Varying r

The effect of increasing the radial pole position r on the frequency response curve is that it becomes sharper and the amplitude increases. We also notice that as the value of r is reduced, the peak frequency changes. We show this by varying the value of r from 0.65 to 0.95. The value of p is kept constant at $p = 1.5321$ ($\theta = 40°$). The plot is shown in Figure 3.21. The Peak frequency is related to the radial pole position r and the angular measure of the pole θ as

$$\cos\theta_{\text{pk}} = 0.25p(r + r^{-1}), \tag{3.25}$$

where θ_{pk} is peak angular frequency. Notice that when $r \to 1$ we get $\theta_{\text{pk}} \to \theta$.

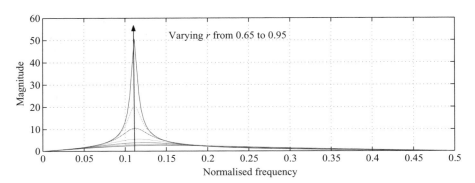

Figure 3.21   Filter response for varying values of r

3.8.2 *Varying p*

Another parameter which can be varied in the filter is p. So we increase the value of p while keeping r constant. From Figure 3.22 we find that the curve shifts towards the right as p increases. The value of p is varied from $p = 1.6180$ ($\theta = 36°$) to $p = -1.6180$ ($\theta = 144°$) keeping the value of r constant at 0.9. Thus we can conclude from the above curves that we can vary the filter characteristics by changing r or by changing p. But if we change the values of r continuously, then the

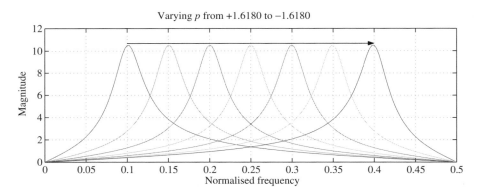

Figure 3.22   Filter response for varying values of p

value of r may exceed one. Under these conditions the system will become unstable, which is not desirable. To keep the system in a stable condition, we vary p. By varying θ we perform a non-linear variation. Hence we choose to vary the value of $p = 2 \cos \theta$. Thus frequency tracking can be done without changing the value of r and by varying p from -2 to 2 and fixing r at an optimum value.

3.8.3 Criteria

We have two parameters, r and p, in the filter. Suppose we need to adapt the filter so its output is maximum. This may sound vague but in a radio receiver we tune, or adjust, the frequency so the listening is best. We now know the sensitivity of the filter – how the parameter affects the characteristics of the filter and to what degree. We define a criteria function $J(p)$, known as the objective function, as follows:

$$J(p) = E(e_k^2)$$
$$= \frac{1}{N}\left(\sum_{i=0}^{N-1} e_i\right), \tag{3.26}$$

where $e_k = x_k - u_k$. Now we plot $J(p)$ against parameter p (Figure 3.23). Notice that a minimum occurs at a single point. It is also essential that most of the time we must know the slope of this function given as $S(p) = \nabla J(p)$.

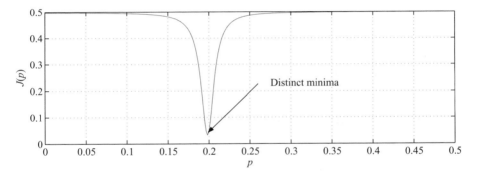

Figure 3.23   Criteria function $J_{(p)}$

This function is depicted in Figure 3.24. Perhaps you are wondering how these functions have been arrived at numerically. We describe this in greater detail in Chapter 6, where we give applications and some problems for research. In brief, the filter is excited by a pure sine wave and $J(p)$ is computed using (3.26). Computing the slope of $J(p)$ requires deeper understanding.

3.8.4 Adaptation

The most popular algorithm for adaptation is the steepest gradient method, also called the *Newton–Raphson method*. For incrementing the parameter p, we use the gradient $\nabla J(p)$, defined as

$$\nabla J(p) = \frac{dJ(p)}{dp}. \tag{3.27}$$

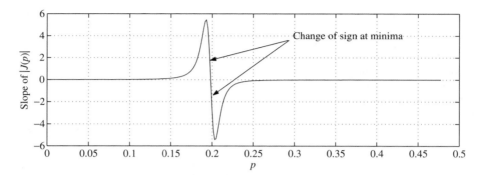

Figure 3.24   The slope of the criteria function is $S(p) = \tilde{N}J(p)$

The parameter incrementing, adaptation or updating is performed as

$$p_{k+1} = p_k - \frac{J(p)}{\nabla J(p)}\mu, \qquad (3.28)$$

$$p_{k+1} = p_k - [J(p)][\nabla J(p)]^{-1}\mu. \qquad (3.29)$$

Here μ is defined as the step size.

3.9 Demodulating via Adaptive Filters

Conventional coherent or non-coherent demodulation techniques for demodulation of binary frequency shift keyed (BFSK) signals do not perform well under low signal-to-noise ratio (SNR) conditions and knowledge of the frequencies poses a limitation. We analyse a demodulation technique in which a sampled version of a BFSK signal is passed through an adaptive digital notch filter. The filter estimates and tracks the instantaneous frequency of the incoming signal in the process of optimising an objective function; a smoothing of the estimated instantaneous frequencies followed by a hard decision gives the demodulated signal.

3.9.1 Demodulation Method

Consider a situation in which a BFSK signal, corrupted by zero-mean additive white noise, is sampled at the receiver. The sampled signal can be expressed as

$$u_k = A\cos(2\pi k f_k) + \gamma_k, \quad k = 0, 1, 2, \ldots, \qquad (3.30)$$

where A is the amplitude ($A > 0$); γ_k is the additive noise, which is a zero-mean white random process with variance σ^2; and f_k is the normalised instantaneous frequency (the actual instantaneous frequency divided by the sampling rate), given by

$$f_k = \begin{cases} f_{\text{mark}} & \text{if the symbol is 1,} \\ f_{\text{space}} & \text{if the symbol is 0.} \end{cases} \tag{3.31}$$

A typical FSK signal and its spectrum are given in Figure 3.25. As in the signal model (3.30), the SNR (in dB) is $10 \log_{10} (A^2/2\sigma^2)$. The signal is passed through

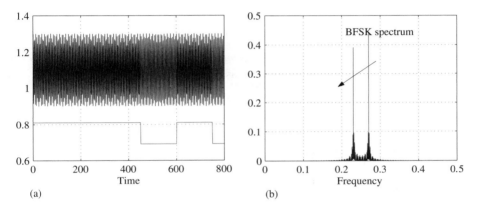

Figure 3.25   BFSK signal in the time domain and the frequency domain

an adaptive digital notch filter that consists of a *digital bandpass filter* (BPF) and a *subtractor*, along with a *sensitivity function estimator* and a *parameter estimator*. The sampled signal sequence $\{u_k\}$ goes through a second-order BPF [6] whose output sequence $\{x_k\}$ is governed by the second-order linear recurrence relation.

$$x_k = rpx_{k-1} - r^2 x_{k-2} + \frac{(1 - r^2)}{2} (u_k - u_{k-2}), \tag{3.32}$$

where $p = 2\cos\theta$ and θ is the normalised centre frequency of the BPF; r is the radial pole position. The subtractor subtracts the input u_k from the BPF output x_k to produce an *error signal*

$$e_k = x_k - u_k. \tag{3.33}$$

The combination of the BPF and the subtractor with input sequence $\{u_k\}$ and output sequence $\{e_k\}$ is a *notch filter*. The parameter estimator estimates the parameter p for every k by trying to minimise the *objective function*

$$J_k(p) = e_k^2, \tag{3.34}$$

applying a gradient descent algorithm on p. Denoting the *sensitivity function* as

$$s_{k-1} = \frac{de_k}{dp} = \frac{dx_k}{dp},$$ (3.35)

it can easily be shown that s_k satisfies the recurrence relation

$$s_k = rps_{k-1} - r^2 s_{k-2} + rx_k.$$ (3.36)

In addition, if v_k represents the mean of $J_k(p)$, then v_k can be computed recursively using an exponentially weighted fading function using the recurrence relation

$$v_k = \lambda v_{k-1} + (1 - \lambda)e_k^2,$$ (3.27)

where λ, $0 < \lambda < 1$, is the *forgetting factor*. Denoting $\hat{p}_k$ as the estimate of p at time index k, we compute $\hat{p}_k$ recursively as

$$\hat{p}_k = \hat{p}_{k-1} - \frac{2e_k s_{k-1}}{v_k} \mu,$$ (3.38)

where μ is the step size. The instantaneous normalised frequency estimate $\hat{f}_k$ is obtained from $\hat{p}_k$ by the formula

$$\hat{f}_k = \frac{1}{2\pi} \cos^{-1}\left(\frac{\hat{p}_k}{2}\right).$$ (3.39)

The *algorithm* for obtaining the sequence $\{\hat{f}_k\}$ of the estimates of the instantaneous normalised frequency from the sampled signal sequence $\{u_k\}$ is as follows.

Choose parameters r, λ, μ and initial values $x_0, x_{-1}, u_0, u_{-1}, s_0, s_{-1}, s_{-2}, v_0, \hat{p}_0$ such that $v_0 > 0$.

For $k = 1, 2, 3, \ldots$

1. $x_k = r\hat{p}_{k-1}x_{k-1} - r^2 x_{k-2} + \dfrac{(1 - r^2)}{2(u_k - u_{k-2})}$
2. $s_k = r\hat{p}_{k-1}s_{k-1} - r^2 s_{k-2} + rx_k$
3. $e_k = x_k - u_k$
4. $v_k = \lambda v_{k-1} + (1 - \lambda)e_k^2$
5. $\hat{p}_k = \hat{p}_{k-1} - \dfrac{2e_k s_{k-1}}{v_k \mu}$
6. $\hat{f}_k = \dfrac{1}{2\pi} \cos^{-1}\left(\dfrac{\hat{p}_k}{2}\right)$

By smoothing the sequence $\{\hat{f}_k\}$ and then performing a hard decision on the smoothed sequence, we obtain the demodulated signal. We have chosen to pass the signal $\{\hat{p}_k\}$ through a hard limiter to reduce the computational burden, which is a requirement in *real-time embedded system* software development.

3.9.2 Step Size μ

In all the methods the choice of μ plays an important role. Various authors [5] have adopted different techniques for the choice of μ. From the value of s_k we can obtain the incremental value of p. A sign change in the slope of a function indicates that a minimum or maximum occurs in between these points. Hence to find the minimum for the function given by (3.34), we need to find the sign change in the slope of $\nabla J_k(p) \cong E\{e_k s_k\}$. Let us define a variable α_k such that

$$\alpha_k = \text{sign}[(E\{e_k s_k\})], \tag{3.40}$$

which may take the value -1 or $+1$ depending on $\nabla J_k(p)$. We change the step size μ to half its original value if we see any change in α_k. In other words, if

$$|\alpha_k - \alpha_{k-1}| = 2, \tag{3.41}$$

it implies that a sign change has occurred and the value of μ changes as

$$\mu_{\text{new}} = \frac{\mu_{\text{old}}}{2}, \tag{3.42}$$

otherwise

$$\mu_{\text{new}} = \mu_{\text{old}}. \tag{3.43}$$

By changing the value of μ for a change in α_k, the value of p converges to the precise value in fewer iterations.

3.9.3 Performance

A signal with additive noise such that the SNR of the signal is 10 dB is implemented for different normalised frequencies of the incoming signal from 0.1 to 0.4 at SNR = 10 dB. A few values are shown in Table 3.1. It shows the performance of the algorithm expressed as the nominal mean and its deviation from the true value as an error. The algorithm is now used to demodulate an FSK by

Table 3.1 Performance of the algorithm

Normalised frequency	Mean	Error
0.2000	0.2000	-1.7762×10^{-6}
0.2100	0.2098	2.4752×10^{-4}
0.2200	0.2200	-2.9443×10^{-5}
0.2300	0.2300	4.4484×10^{-6}

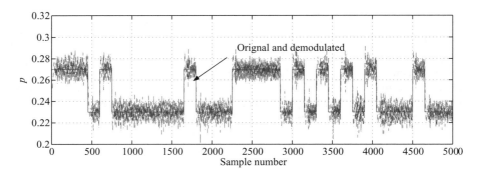

Figure 3.26   Demodulated BFSK signal p

tracking the mark and space frequencies. An implementation for an FSK signal with SNR $= 10\,$dB is shown in Figure (3.26).

3.10 Phase Shift via Adaptive Filter

In the application of detection, array processing and direction finding, we need to take the given narrowband signal (here we have chosen a sinusoidal wave) and shift it by a given value. The following example generates a $\pi/2$ phase shift by adjusting the coefficient of the first-order all-pass filter. Consider the given signal u_k as

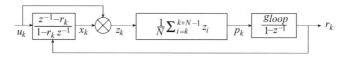

Figure 3.27   Adaptive phase shifting

$u_k = \cos(2\pi fk + \phi)$ where f is the normalised frequency and ϕ is an arbitrary phase value. Figure 3.27 depicts the macro view of the filter.

$$u_k = \cos(2\pi fk + \phi), \tag{3.44}$$

$$x_k = r_k x_{k-1} - r_k u_k + u_{k-1}, \tag{3.45}$$

$$z_k = u_k x_k \quad \text{and} \quad e_k = \left(\frac{1}{N}\right)[z_k - z_{k-N}], \tag{3.46}$$

$$p_k = p_{k-1} + e_k, \tag{3.47}$$

$$r_k = r_{k-1} - g_{\text{loop}} p_k \quad \text{if} \quad |r_k| \geq 0.95 \quad \text{then} \quad r_k = 0.95. \tag{3.48}$$

The above filter is implemented for $N = 40, g_{\text{loop}} = 0.01$ and $1/f = 77$ with $\phi = \pi/20$ and an initial value of $r_k = 0.2$. Figure 3.28 shows the phase shift.

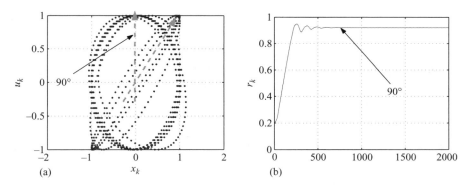

Figure 3.28   Phase shifting by an adaptive filter

3.11 Inverse Problems

Inverse problems are classified under parameter estimation. Originally the playground of *statisticians*, people from many disciplines have now entered this field. It is best used for many applications in DSP.

In general, it is a complex problem to find the difference equation or the digital filter that yields the given output for a given input, and a solution cannot be assured. Often a unique solution is impossible. Each problem must be treated individually. However, we will describe a generic method and enumerate the shortcomings. This is a fairly big area, so it is not possible to do more than give a few relevent pointers. Chapter 6 gives some applications using this idea.

There are two problems in this area: non-availability of knowledge of the filter order, and the presence of noise in measurements. These make the solution difficult.

3.11.1 Model Order Problem

Let us consider two digital filters:

$$x_k = 1.7x_{k-1} - 1.53x_{k-2} + 0.648x_{k-3} + u_k + 0.6u_{k-1}, \tag{3.49}$$

$$x_k = 2.17x_{k-1} - 2.413x_{k-2} + 1.474x_{k-3} - 0.365x_{k-4} + u_k. \tag{3.50}$$

Look at Figure 3.29 (a) and (b). They represent the frequency responses of the digital filters in (3.49) and (3.50). We have superimposed the two responses in Figure 3.29(b). Notice that there is a good match in an engineering sense.

Multiple solutions make life more complicated in inverse problems. The reason for this is that both filters have different phase responses or could have a pole-zero cancellation. One of the best criteria for fixing the model order is by Akaike, known as the Akaike information criterion (AIC). Given an input sequence $\{u_k\}$ and output sequence $\{y_k\}$, using AIC and human judgement, by looking at the spectrum of $\{y_k\}$, we can arrive at a fairly accurate estimate for p and q in (2.15), where p is the

ARMA(3,2)
(a)
ARMA(4,1)
and ARMA(3,2)
(b)

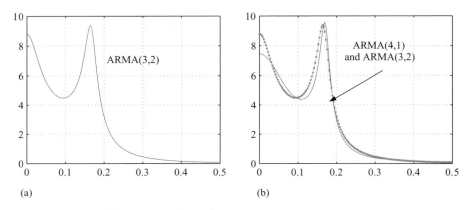

Figure 3.29 Two different filters give the same frequency response

order of the numerator polynomial and q is the order of the denominator polynomial.

3.11.2 Estimating Filter Coefficients

Refer to (1.24) to (1.26) for the definition of the vector $\mathbf{p} = [a_1, a_2, \ldots, a_p, b_1, \ldots, b_q]$ which defines the digital filter. We also have another vector $\mathbf{z}_k^t = [y_{k-1}, \ldots, y_{k-p}, u_k, \ldots, u_{k-q}]$. Then we construct an equation representing the given data as $y_k = \mathbf{z}_k \mathbf{p}^t$. This is exactly the same as the set of overdetermined simultaneous equations given in Chapter 2 and the solution has the form of (2.35). Using (2.36) and (2.33), we arrive at the solution for $\mathbf{p}$ as

$$\mathbf{p} = \left(\sum_{k=1}^{N} \mathbf{z}_k \mathbf{z}_k^t \right)^{-1} \left(\sum_{i=1}^{N} y_k \mathbf{z}_k^t \right). \tag{3.51}$$

There are many methods centred around this solution and they are shown in Figure 3.29A.

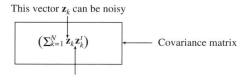

This vector $\mathbf{z}_k$ can be noisy
$\left(\sum_{k=1}^{N} \mathbf{z}_k \mathbf{z}_k^t\right)$
Covariance matrix
A noise-free or independent source of the $\mathbf{z}_k^t$-like vector is the *key*

Figure 3.29A There are many methods centred around solution (3.51)

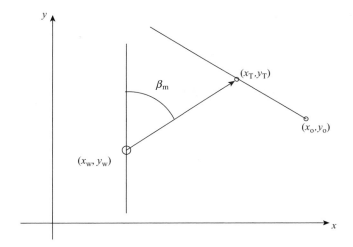

Figure 3.30   Tracking a moving object

3.11.3 Target Tracking as an Inverse Problem

In real life, formulation of a problem constitutes 50% of the solution. Let us consider an object moving with a uniform speed in a straight line (Figure 3.30). At $t = 0$, or in the discrete-time domain at $k = 0$, we assume it has a value of (x_0, y_0). We observe the target coordinates (x_T, y_T) from a movable watching place at (x_w, y_w). The angle with respect to the true north is $\beta_m = tan^{-1}[(x_T - x_w)/(y_T - y_w)]$.

3.11.3.1 Problem Formulation

We can recast the problem as

$$\frac{\sin \beta_m}{\cos \beta_m} = \frac{x_T - x_w}{y_T - y_w}.$$

We rewrite the equation as

$$x_T \cos \beta_m - y_T \sin \beta_m = x_w \cos \beta_m - y_w \sin \beta_m. \qquad (3.52)$$

Under the assumption that the target is travelling at uniform speed in a constant direction, we get $x_T = x_0 + k\,\delta t\,v_x$ and $y_T = y_0 + k\,\delta t\,v_y$, which leads to

$$x_0 \cos \beta_m - y_0 \sin \beta_m + k\,\delta t \cos \beta_m v_x - k\,\delta t \sin \beta_m v_y = x_w \cos \beta_m - y_w \sin \beta_m. \qquad (3.53)$$

To bring this into standard form, we put

$$\mathbf{u}^t = [\cos \beta_m - \sin \beta_m \, k\delta t \, \cos \beta_m - k \, \delta t \, \sin \beta_m]$$

and $\mathbf{b}' = [x_o \; y_o \; v_x \; v_y]$. Representing the output as a scalar $y_k = x_w \cos \beta_m - y_w \sin \beta_m$, we get

$$\mathbf{u}'\mathbf{b} = y_k. \tag{3.54}$$

3.11.3.2 Pseudolinear Solution

The above formulation is a pseudo MA filter. This is an inverse problem of finding $\mathbf{b}$ given y_k and $\mathbf{u}$. The solution can be of block type or recursive type. For simplicity we show a block solution which is given as $\mathbf{b} = \left(\sum_{i=1}^{N} \mathbf{u}_i \mathbf{u}_i^t \right)^{-1} \sum_{i=1}^{N} (\mathbf{u}_i y_i)$; it is depicted in Figure 3.31(a). It shows the motion of the watcher or observer, the motion of the target and the estimated motion of the target. Figure 3.31(b) shows the angle (β_m) measurements.

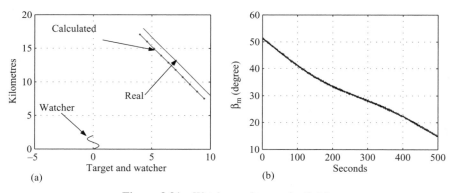

Figure 3.31   Watcher and target in (3.54)

A careful look at Figure 3.31(a) reveals that the estimate is biased, even though the direction is correct. This is because we have introduced a small amount of noise in the measured angle β_m. In fact, it is a very small bias but still significant. A considerable amount of published work is available in the area known as target motion analysis (TMA). There are many methods to remove this bias, such as instrumental variable (IV) methods, maximum likelihood (ML) methods and IV approximate ML (IVAML) estimators. There are many associated problems still to be researched.

3.12 Kalman Filter

The output of a filter is completely characterised by its coefficients and the initial conditions (Figure 3.31A).

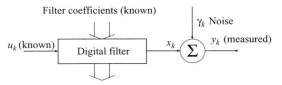

State of the filter (to be estimated)

Figure 3.31A Kalman filter: the output is completely characterised by the filter coefficients and the initial conditions

For a simple explantion, consider (3.16), where the infinite sequence is generated by a 3-tuple $\{1.8, 1, 0\}$ in which there are two components: coefficients, known as parameters; and the other values at various places inside the filter before the delay elements, known as the state of the filter.

The output is generally a linear combination of these states. The aim of the Kalman filter is to accurately estimate these states by measuring the noisy output. The following examples illustrate the approach and its application.

3.12.1 Estimating the Rate

Estimating the rate occurs in radar, tracking and control. Let us consider a sequence of angular measurements $\{\theta_1, \theta_2, \ldots, \theta_N\}$ over a relatively small window.

$$\theta \text{ noisy} \longrightarrow \boxed{\text{Piecewise linear model}} \longrightarrow \begin{array}{c} \hat{\theta} \\ \dot{\theta} \end{array}$$

It is required to find the slope or the rate. The problem looks simple, but using conventional Newton forward or backward differences produces noisy rates. Consider the simple plot of θ_k in Figure 3.32.

Now we place the x and y axes at the desired point, as shown in Figure 3.32. Note that we have moved only the y-axis to the desired place. Then we have $\theta = mk \, \delta t + c$. Here we need to estimate the value of m. This choice of coordinate frame ensures that $k = 0$ at all times. We can write

$$\begin{pmatrix} \theta_1 \\ \theta_2 \\ \vdots \\ \theta_N \end{pmatrix} = \begin{pmatrix} 0 & 1 \\ \delta t & 1 \\ \vdots & \vdots \\ (N-1)\delta t & 1 \end{pmatrix} \begin{pmatrix} m \\ c \end{pmatrix} \quad \text{or } \Theta = \mathbf{A}\mathbf{p}. \qquad (3.55)$$

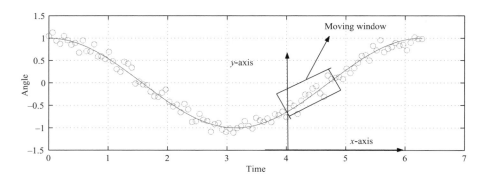

Figure 3.32   Linear regression over a rectangular moving window

Now the solution for such overdetermined equations in a least-squares sense is $\mathbf{p} = (\mathbf{A}^t\mathbf{A})^{-1}\mathbf{A}^t\Theta$, where our interest is $p(1) = m$. We can derive a closed-form solution as

$$\binom{m}{c} = \begin{pmatrix} \dfrac{12}{N(N^2 - 1)} & \dfrac{-6}{N(N - 1)\delta t} \\ \dfrac{-6}{N(N - 1)\delta t} & \dfrac{2(2N + 1)}{N(N - 1)} \end{pmatrix} \mathbf{A}^t\Theta. \tag{3.56}$$

Even though (3.56) looks very complicated, it is only a pair of FIR filters with their impulse responses given as $h_k^A = \{2, 4, 6, 8, \ldots\}$ and $h_k^B = \{1, 1, 1, 1, \ldots\}$. For a six-point moving linear regressor, the rate is given as

$$\dot{\theta}_k = \frac{1}{\delta t}\left\{(6 + 1)\left[\sum_{i=0}^{6-1}\theta_{k-i}\right] - \left[\sum_{i=0}^{6-1}2(i + 1)\theta_{k-i}\right]\right\}\left(\frac{1}{6^2 - 1}\right). \tag{3.57}$$

This is shown in Figure 3.33, which is a SIMULINK block diagram. In fact, this is a Kalman filter or a state estimator.

3.12.2 Fitting a Sine Curve

There are many situations where you know the frequency, but unfortunately, by the time you get the signal, it is corrupted and of the form $y_k = \sin(2\pi f k) + n_k$, where n_k is a zero-mean uncorrelated noise. The problem is to get back the original signal. Immediately what strikes our mind is a high-Q filter. Unfortunately, stability is an issue for these filters. In situations of this type, we use model-based filtering. Consider an AR process representing an oscillator (poles on the unit circle):

$$x_k = px_{k-1} - x_{k-2} \quad \text{where} \quad -2 \le p \le 2. \tag{3.58}$$

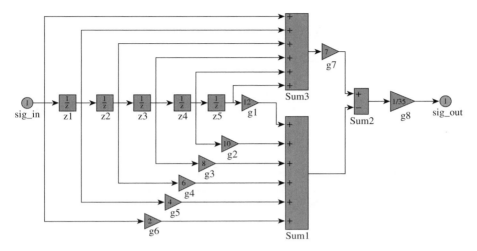

Figure 3.33 Moving linear regression

The parameter p has a direct relationship with the frequency. We can see that the initial conditions of this difference equation uniquely determine the amplitude and phase of the desired noise-free sine wave. Now we aim at finding out the initial conditions of this difference equation that generate the best-fit sine wave matching the noisy generated sine wave (Figure 3.34). To formulate this problem, we choose to write it as

$$\begin{pmatrix} x_k \\ x_{k-1} \end{pmatrix} = \begin{pmatrix} p & -1 \\ 1 & 0 \end{pmatrix} \begin{pmatrix} x_{k-1} \\ x_{k-2} \end{pmatrix} \quad \text{or} \quad \mathbf{x}_k = \Theta \mathbf{x}_{k-1}. \qquad (3.59)$$

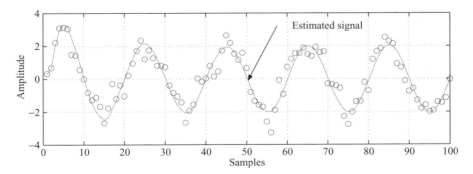

Figure 3.34 Fitting a sine curve to a noisy signal

We also define a vector $\mathbf{c}' = (1\ 0)$. Then the noisy measured signal will be $y_k = \mathbf{c}'\Theta\mathbf{x}_{k-1} + n_k$. Now let the unknown initial condition vector be $\mathbf{x}_0$. Then we have

$$
\begin{pmatrix} y_1 \\ y_2 \\ \vdots \\ y_N \end{pmatrix} = \begin{pmatrix} \mathbf{c}'\Theta \\ \mathbf{c}'\Theta^2 \\ \vdots \\ \mathbf{c}'\Theta^N \end{pmatrix} \mathbf{x}_0 + \begin{pmatrix} n_1 \\ n_2 \\ \vdots \\ n_N \end{pmatrix}, \tag{3.60}
$$

which is of the form $\mathbf{y} = \mathbf{A}\mathbf{x}_0 + \mathbf{n}$. Now we know that the least-squares solution for the initial conditions of the oscillator is $\mathbf{x}_0 = (\mathbf{A}'\mathbf{A})^{-1}\mathbf{A}'\mathbf{y}$.

We do not need to do these computational gyrations. Fortunately, we can use a fine numerically efficient recursive method known as a Kalman filter. The following set of equations provide a recursive solution and update the desired estimate of the signal $\hat{y}_k$ on receipt of new data:

$$
\begin{aligned}
\hat{\mathbf{x}} &= \Theta\mathbf{x}_k, \\
e_k &= \mathbf{c}'\hat{\mathbf{x}} - y_k, \\
\hat{\mathbf{P}} &= \Theta\mathbf{P}_k\Theta, \\
\mathbf{x}_{k+1} &= \hat{\mathbf{x}} - \frac{\hat{\mathbf{P}}\mathbf{c}'}{1 + \mathbf{c}\hat{\mathbf{P}}\mathbf{c}'} e_k, \\
\mathbf{P}_{k+1} &= \hat{\mathbf{P}} - \frac{\hat{\mathbf{P}}\mathbf{c}'\mathbf{c}\hat{\mathbf{P}}}{1 + \mathbf{c}\hat{\mathbf{P}}\mathbf{c}'},
\end{aligned} \tag{3.61}
$$

$\hat{y}_k = \mathbf{c}'\mathbf{x}_{k+1}$ is the desired filtered signal.

We recommend you to look at the above as a numerical method for implementing

$$
y_k = \mathbf{c}'\Theta^k(\mathbf{A}'\mathbf{A})^{-1}\mathbf{A}'\mathbf{y}. \tag{3.62}
$$

The choice of $\mathbf{P}_0 = 10^{-3}\mathbf{I}$ is good and we can start with any arbitrary value of $\mathbf{x}_k$. Programs given in the appendix will throw more light on this.

3.12.3 Sampling in Space

Consider an antenna consisting of the uniform linear array (ULA) shown in Figure 3.35. We assume a radiating narrowband source transmitting in all directions. The signal as seen by the ULA at a given instant of time is a sine wave or part of it, depending on the interelement spacing. For this reason it is called sampling in space. The set of signals measured by the ULA is noisy and given as $y_n = x_n + \gamma_n$, where γ_n is a zero-mean uncorrelated random sequence.

This example aims at obtaining a set of coefficients for a pair of FIR filters which satisfy the property that the output of the filter pair acts as a state estimate for a

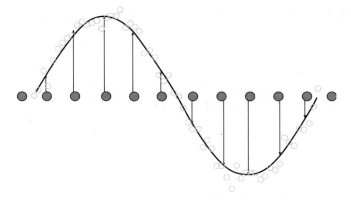

Figure 3.35   Spatial wave on a ULA of L measurements $y_{1,n}$, K, $y_{L,n}$

sinusoidal oscillator. These coefficients are then modified and applied to a *direction finding* (DF) system using a ULA.

3.12.3.1 Angular Frequency Estimation

Consider an AR process given by

$$x_{n+1} = 2 \cos(2\pi f)x_n - x_{n-1},$$

representing an oscillator with *normalised angular frequency* $\theta \triangleq 2\pi f$. By defining

$$\mathbf{X}_n(\theta) \triangleq [x_{n+1}, \, x_n]^T, \qquad \Phi(\theta) \triangleq \begin{bmatrix} 2\cos\theta & -1 \\ 1 & 0 \end{bmatrix},$$

this can be rewritten as $\mathbf{X}_n(\theta) = \Phi(\theta)\mathbf{X}_{n-1}(\theta)$, where $\mathbf{X}_n(\theta)$ is the *state vector* and (θ) is the *state transition matrix*. For the nth snapshot $(n = 1, \ldots, N)$ let L measurements $y_{1,n}, \ldots, y_{L,n}$ of $x_n, \ldots, x_{n+L-1}$ be written as

$$y_{k,n}(\theta) = \mathbf{c}^T \mathbf{X}_n(\theta) + \gamma_{k,n}, \quad k = 1, \ldots, L, \tag{3.63}$$

where $\mathbf{c} \triangleq [0, \, 1]^T$, and $\gamma_{k,n}$ is additive noise. An $L \times 1$ *measurement vector*

$$\mathbf{Y}_n(\theta) \triangleq [y_{1,n}, \ldots, y_{L,n}]^T$$

can be expressed using (3.63) as

$$\mathbf{Y}_n(\theta) = \mathbf{A}(\theta)\mathbf{X}_n(\theta) + \Gamma_n, \tag{3.64}$$

where Γ_n is the additive noise vector, and

$$\mathbf{A}(\theta) = \begin{bmatrix} 0 & 1 & \frac{\sin(2\theta)}{\sin\theta} & \cdots & \frac{\sin((L-1)\theta)}{\sin\theta} \\ 1 & 0 & -1 & \cdots & -\frac{\sin((L-2)\theta)}{\sin\theta} \end{bmatrix}^T.$$

The *least squares estimate* (LSE) of $\mathbf{X}_n(\theta)$ in (3.64) is

$$\hat{\mathbf{X}}_n(\theta) = [\mathbf{A}^T(\theta)\mathbf{A}(\theta)]^{-1}\mathbf{A}^T(\theta)\mathbf{Y}_n(\theta).$$

Each row of the weighting matrix $[\mathbf{A}^T(\theta)\mathbf{A}(\theta)]^{-1}\mathbf{A}^T(\theta)$ represents an FIR filter of L coefficients.

Consider a measurement vector $\mathbf{Y}_n(\omega)$ for an oscillation at ω. If $\mathbf{X}_n(\omega)$ is computed using $\Phi(\theta)$, then its LSE is given by $\hat{\mathbf{X}}_n(\theta, \omega) = [\mathbf{A}^T(\theta)\mathbf{A}(\theta)]^{-1}\mathbf{A}^T(\theta)\mathbf{Y}_n(\omega)$. The estimate of $\mathbf{Y}_n(\omega)$ is then

$$\hat{\mathbf{Y}}_n(\theta, \omega) = \mathbf{A}(\theta)\hat{\mathbf{X}}_n(\theta, \omega) = \mathbf{P}_A(\theta)\mathbf{Y}_n(\omega).$$

3.13 Summary

This chapter described ways to specify a given filter. We presented a variety of popular filters and their characteristics. We looked at IIR filters with real coefficients and truncated IIR filters for use in FIR designs. We also described a bank of filters with complex coefficients pointing to discrete Fourier transforms. We gave a simple presentation of adaptive filters and inverse problems having practical significance. We demonstrated an application of BFSK demodulation and adaptive phase shifting. And we looked at a practical inverse problem from target tracking. We added an important dimension by giving Kalman filter applications. Then we concluded this big topic with an application from array processing.

References

1. R. W. Hamming, *Digital Filters*, pp. 4–5. Englewood Cliffs, NJ: Prentice Hall, 1980.
2. R. D. Strum and D. E. Kirk, *First Principles of Discrete Systems and Digital Signal Processing*. Reading MA: Addison-Wesley, 1989.
3. K. V. Rangarao, Adaptive digital notch filtering. Master's thesis, Naval Postgraduate School, Monterey CA, September 1991.
4. G. C. Goodwin and K. S. Sun, *Adaptive Filtering, Prediction and Control*. Englewood Cliffs NJ: Prentice Hall, 1984.
5. D. M. Himmelblau *Applied Non-linear Programming*. New York: McGraw-Hill, 1972.
6. K. Kadambari, K. V. Rangarao and R. K. Mallik, Demodulation of BFSK signals by adaptive digital notch filtering. In *proceedings of the IEEE International Conference on Personal Wireless Communications*, Hyderabad, India, December 2000, pp. 217–19.

4

Fourier Transform and Signal Spectrum

Any book on DSP is not complete without a detailed discussion of the Fourier transform. There are many publications in this area, but Brigham [1] is probably the best. The fast Fourier transform (FFT) is the numerical method [2] for computing the discrete Fourier transform (DFT) of a given time series. It exploits the periodicity properties of the function $e^{j2\pi k}$ where k is an integer.

4.1 Heterodyne Spectrum Analyser

Analysing the spectrum using the superheterodyning principle is very popular due to the simple nature of its implementation. The idea is to create a relative motion between the bandpass filter (BPF) and the incoming signal spectrum:

$$X(f) \triangleq \int_{-\infty}^{\infty} x(t)e^{-j2\pi ft}\,dt. \tag{4.1}$$

Equation (4.1) is shown in Figure 4.1 to give a better insight. The digitised signal s_k is multiplied with a frequency-swept signal x_k. Note that we need to sweep using a staircase function[1] not a ramp function. We have presented the results at SNR $= 0$ dB.

4.1.1 Spectrum Analyser

The signal x_k is defined as

$$x_k = \cos(2\pi f_k k) \text{ where } f_k = \text{staircase}(N, \delta f). \tag{4.2}$$

---

[1]This function permits the BPF to settle down for each *step* change δf, as shown in Figure 4.2.

Digital Signal Processing: A Practitioner's Approach   K. V. Rangarao and R. K. Mallik
© 2005 John Wiley & Sons, Ltd

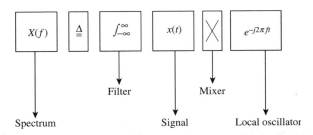

Figure 4.1   An engineer's view of (4.1)

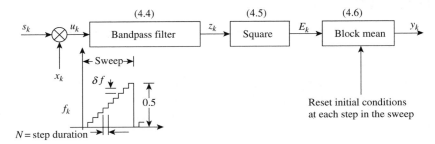

Figure 4.2   Digital spectrum analyser

The frequency f_k is a staircase function totally defined by the step duration N and the step height $\delta f = 0.5/n$, where n is the number of samples of the spectrum.

A typical block diagram of this method is shown in Figure 4.2 along with the *staircase* function. It has three components: a BPF, a square function and a synchronised block averager. This block averager is discussed in Chapter 3 and is shown as a recursive filter. We can depict Figure 4.2 in the form of digital filters:

$$u_k = s_k x_k, \tag{4.3}$$

$$z_k = rp z_{k-1} - r^2 z_{k-2} + \left(\frac{1-r^2}{2}\right)(u_k - u_{k-2}), \tag{4.4}$$

$$E_k = z_k^2, \tag{4.5}$$

$$y_k = \left(\frac{1}{1+\frac{1}{N}}\right) y_{k-1} + \left(\frac{1}{N+1}\right) E_k, \tag{4.6}$$

where the value of p lies between -2 and $+2$. Equation (4.6) is for obtaining the mean recursively, as described in (3.3). The initial conditions need to be reset at each step change.

4.1.1.1 Output

A spectrum analyser is implemented using $r = 0.98$ and $p = 0$ for the BPF. The staircase function is used with $N = 150$ and $n = 200$.

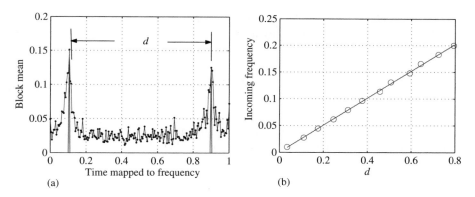

Figure 4.3 Output of a block mean filter at an SNR of 0 dB

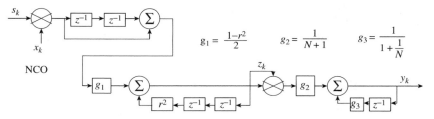

Figure 4.4 Hardware realisation of a spectrum analyser

Figure 4.4 shows a hardware realisation of the filters in (4.3) to (4.6). Figure 4.3(a) shows the output y_k (4.6) for a specific value of s_k. Figure 4.3(b) shows the distance between the peaks versus the input frequency of the signal s_k; there is a perfect linear relation. In reality, we don't have to sweep till 0.5 Hz and it will suffice if we sweep till 0.25 Hz and find the values of y_k.

4.2 Discrete Fourier Transform

The DFT relation is derived from the continuous FT of a signal. Consider a signal $x(t)$ and its FT as

$$X(f) \triangleq \int_{-\infty}^{\infty} x(t)e^{-j2\pi ft}\, \mathrm{d}t. \tag{4.7}$$

If the signal is such that $x(t) = 0$ for $t < 0$, then the FT takes the form

$$X(f) = \int_{0}^{\infty} x(t)e^{-j2\pi ft}\, \mathrm{d}t. \tag{4.8}$$

In (4.8) we replace the integral with a summation ($\int \rightarrow \sum$), the continuous variable t with the discrete variable n ($t \rightarrow n$), and f with k ($f \rightarrow k$). We get the DFT relation for sequence x_n of length N:

$$X_k = \sum_{n=0}^{N-1} x_n e^{-j\,\delta\theta\,nk} \quad \text{where} \quad \delta\theta = \frac{2\pi}{N}. \tag{4.9}$$

It is very convenient and mathematically tractable to use the notation $W_N = e^{j2\pi/N}$ and rewrite (4.9) as

$$X_k = \sum_{n=0}^{N-1} x_n W_N^{nk}. \tag{4.10}$$

The complex quantity W_N is a constant for a given value of N. Thinking of W_N as $\cos(\delta\theta) + j\sin(\delta\theta)$ will probably help you to understand it. Let us look at the transformation $W_N^2 = \left(e^{j2\pi/N}\right)^2$. This can be written as $W_N^2 = e^{j2\pi/(N/2)}$, which is the same as $W_{N/2}$. It may become clearer by looking at a numerical example such as $W_{16}^2 = W_8$.

4.3 Decimating the Given Sequence

Equation (4.10) can also be written in matrix form for N points:

$$\mathbf{S} = (\mathbf{W}_N)\mathbf{x} \tag{4.11}$$

where $\mathbf{S}$ is a vector with components $X_1, \ldots, X_{N-1}$.

Each row of the matrix is an FIR filter with complex coefficients (Figure 4.5), and coefficients are sampled complex cosine and sine waves. Each row can be viewed as an oscillator. The frequency of the sinewave in each row increases linearly as we move down the matrix from top to bottom. This idea was implicitly demonstrated in Chapter 3.

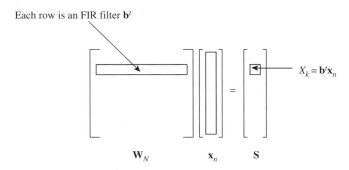

Figure 4.5   Discrete Fourier transform: understanding (4.11)

4.3.1 Sliding DFT

The transfer function of the kth row filter can also be written as an IIR filter

$$H_k(z) = \frac{1 - z^{-N}}{1 - W_N^{-k} z^{-1}} \quad \text{for} \quad k = 0, 1, \ldots, N - 1. \tag{4.12}$$

The pole $z = W_N^{-k}$ of $H_k(z)$ is cancelled by one of its zeros, resulting in an FIR filter. For $N = 8$ we get

$$\frac{(1 - W_8^4 z^{-1})(1 - W_8^3 z^{-1})(1 - W_8^{-3} z^{-1})(1 - W_8^2 z^{-1})(1 - W_8^{-2} z^{-1})(1 - W_8^0 z^{-1})(1 - W_8^1 z^{-1})(1 - W_8^{-1} z^{-1})}{1 - W_8^{-k} z^{-1}},$$

$$\tag{4.13}$$

where $W_8 = e^{-j\pi/4}$. This example shows the pole–zero cancellation. Each kth row k can also be written in the form of a difference equation:

$$s_n = W_N^{-k} s_{n-1} + \Delta x, \tag{4.14}$$

where $\Delta x = (x_n - x_{n-N+1})$ can be treated as an input to the filter.

4.4 Fast Fourier Transform

Matrix multiplication in (4.11) can be done very efficiently. Since coefficients in the matrix $\mathbf{W}_N$ are periodic, we can arrive at a much more efficient method of computing. The given sequence can be transformed to the frequency domain by multiplying with an $N \times N$ matrix. Direct computing involves sizeable computations and is[2] $OO(N^2)$ Consider a 16-point sequence of numbers for which the DFT equation is

$$X_k = \sum_{n=0}^{15} x_n W_{16}^{nk}. \tag{4.15}$$

We illustrate decimation in time and frequency with a numerical example. Note that the value of N is such that $N = 2^p$ where p is an integer.

4.4.1 Windowing Effect

A finite observation period is limited for many physical reasons and usually produces an aberration in the computed spectrum. Consider $x_k = \cos(2\pi f k)$;

---

[2] OO is to be read as *order of*.

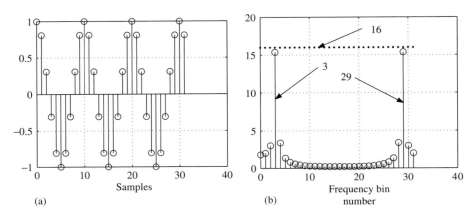

Figure 4.6 Windowing effect

mathematically, the DFT of this sequence must result in a unit sample at $\pm f$. Numerically, let us consider $f = 1/10$ implying that we have chosen 10 samples/ cycle, as shown in Figure 4.6(a). We have limited our observation to 32 samples. Moving from[3] Figure 4.6(a) to Figure 4.6(b) can be depicted using (4.11):

$$\underbrace{\text{Fig 4.6(a)} \rightarrow \text{Fig 4.6(b)}}_{(\mathbf{W}_{32})\mathbf{x}\rightarrow\mathbf{S}}.$$

Note that Figure 4.6(b) corresponds to $|\mathbf{S}|$.

Ideally, at bins 3 and 29 we should have obtained a value of $32/2 = 16$ and the rest of the places should have been zero, but in reality we obtained something different, as shown in Figure 4.6(b). This spectral distortion is explained in Section 1.4. It is attributed to the fact that a rectangular window has the form $\frac{\sin(x)}{x}$ and its associated leakage in the sidelobes. To mitigate the problem of limited observations, we multiply the observations point by point with another sequence $\mathbf{w}$ as in Figure 4.7(a). Operationally we write it as

$$\underbrace{\text{Fig 4.6(a)} \rightarrow \text{Fig 4.7(a)}}_{\mathbf{w}(.\times)\mathbf{x}\rightarrow\mathbf{z}}.$$

The operation $(.\times)$ denotes point-by-point multiplication of two sequences. The non-rectangular windowed spectrum is shown in Figure 4.7(b). This meets one of the criteria that the value of the spectrum be zero other than at $\pm f$ in an approximate sense.

---

[3]Actually we are moving from the *time* domain to the *frequency* domain.

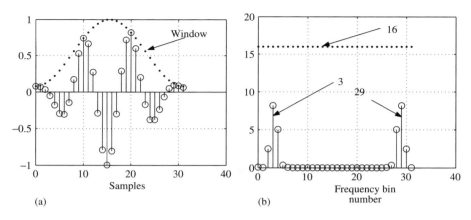

Figure 4.7   Non-rectangular windowing effect

4.4.2 Frequency Resolution

The given frequency in -0.5 to 0.5 is divided into 32 divisions, giving a resolution of $\delta f = 1/N$. The greater the observation time, the better the resolution in the spectrum. This gives us an opportunity to look at the DFT as a bank of filters separated at δf. Due to conjugate symmetry, the useful range is 0 to 15 bins.

4.4.3 Decimation in Time

We could also split the input sequence into odd and even sequences then compute the DFT. We can write (4.15) in a different way as

$$X_k = \sum_{n=\text{even}}^{14} x_n W_{16}^{nk} + \sum_{n=\text{odd}}^{15} x_n W_{16}^{nk}. \tag{4.16}$$

For the even case, n takes the form $2m$; for the odd case, n takes the form $2m + 1$. With this in mind we write (4.16) as

$$X_k = \sum_{m=0}^{7} x_{\text{even}} W_{16}^{2mk} + \sum_{m=0}^{7} x_{\text{odd}} W_{16}^{(2m+1)k}$$
$$= \left(\sum_{m=0}^{7} x_{\text{even}} W_8^{mk} \right) + W_{16}^{k} \left(\sum_{m=0}^{7} x_{\text{odd}} W_8^{mk} \right). \tag{4.17}$$

There are two steps in obtaining a 16-point DFT using an 8-point DFT. First, we divide the time sequence as *odd* and *even* or as two sets of alternating points. We compute the 8-point DFT of the two sequences, shown as step A in Figure 4.8.

Step A   **Step B**

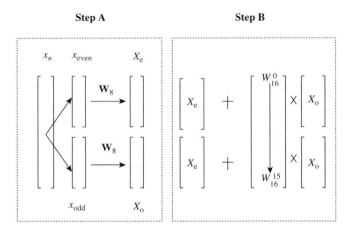

Figure 4.8   Decimation in time

We replicate the odd sequence DFT to make a 16-point sequence and multiply point by point with a factor $\mathbf{W}_{16}^k$. We generate another 16-point sequence by replicating the even DFT sequence. By adding together both sequences we will get the desired 16-point DFT, shown as step B in Figure 4.8. This process is *recursive*, hence it leads to very efficient DFT computation.

4.4.4 Decimation in Frequency

For decimation in frequency, we divide the sequence into *two* halves, premultiply with weights known as *twiddle* factors and compute the DFT (Figure 4.9). We rewrite (4.15) as

$$X_k = \sum_{n=0}^{7} x_n W_{16}^{nk} + \sum_{n=8}^{15} x_n W_{16}^{nk}. \tag{4.18}$$

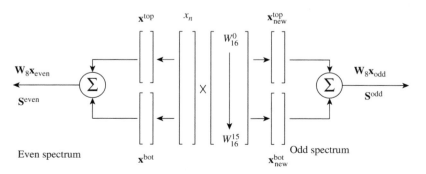

Figure 4.9   Decimation in frequency

In (4.18) when k is even, we can write $k = 2m$ for $m = 0$ to 7:

$$X_k^{even} = \sum_{n=0}^{7} x_n W_{16}^{n2m} + \sum_{n=8}^{15} x_n W_{16}^{n2m}$$

$$= \sum_{n=0}^{7} x_n W_8^{nm} + \sum_{n=8}^{15} x_n W_8^{nm} \tag{4.19}$$

$$= \left(\sum_{m=0}^{7} x_m W_8^{nm} \right) + \left(\sum_{m=0}^{7} x_{8+m} W_8^{nm} \right). \tag{4.20}$$

If we divide the sequence x_n into two halves as $\mathbf{x}^{top}$ and $\mathbf{x}^{bot}$, then (4.20) has the structure

$$\mathbf{S}^{even} = \left[\mathbf{W}_8 \mathbf{x}^{top} + \mathbf{W}_8 \mathbf{x}^{bot} \right]$$

$$= [\mathbf{W}_8] \left[\mathbf{x}^{top} + \mathbf{x}^{bot} \right] \tag{4.21}$$

Note that even values of the spectrum are obtained by dividing the given sequence into two halves and adding them. Then we take the DFT of this new sequence, which is only half the original sequence. In (4.18) when k is odd, we can write $k = 2m + 1$ for $m = 0$ to 7:

$$X_k^{odd} = \sum_{n=0}^{7} x_n W_{16}^{n(2m+1)} + \sum_{n=8}^{15} x_n W_{16}^{n(2m+1)}$$

$$= \sum_{n=0}^{7} \left[x_n W_{16}^{n} \right] W_8^{nm} + \sum_{n=8}^{15} \left[x_n W_{16}^{n} \right] W_8^{nm} \tag{4.22}$$

$$= \left(\sum_{m=0}^{7} \left\{ x_m W_{16}^{m} \right\} W_8^{nm} \right) + \left(\sum_{m=0}^{7} \left\{ x_{m+8} W_{16}^{m+8} \right\} W_8^{nm} \right) \tag{4.23}$$

Looking at (4.23) in detail, we have created *two* new sequences by doing a point-by-point multiplication with another constant sequence given as

$$\mathbf{x}_{new}^{top} = [\ldots x_m W_{16}^{i} \ldots] \quad \text{for} \quad i = 0 \text{ to } 7, \mathbf{x}_{new}^{bot} = [\ldots x_{m+i} W_{16}^{m+i} \ldots] \quad \text{for} \quad i = 0 \text{ to } 7. \tag{4.24}$$

With this notation the remaining part of the spectrum is given as

$$\mathbf{S}^{odd} = [\mathbf{W}_8] \left[\left(\mathbf{x}_{new}^{top} + \mathbf{x}_{new}^{bot} \right) \right]. \tag{4.25}$$

Why are we complicating things with so many variables? The reason is that it gives us some structure. We summarise the equations as follows:

1. Given the sequence we generate one more extra sequence by multiplying term by term with a complex quantity $W_N^i = \cos(n\,\delta\theta) + j\sin(n\,\delta\theta)$
2. These two sequences are divided into two halves.

3. Add corresponding halves to get *two* half-sequences.
4. Compute the DFT of these two half-sequences.

4.4.5 Computing Effort Estimate

Consider a real sequence of length N. Computing the DFT is equivalent to multiplying an $N \times N$ complex matrix with an N-vector. Each row-by-column multiplication needs $2N$ multiplications and $2N$ additions. Assuming there is not much difference between multiplication and addition (true in the case of floating point), the total number of operations per row is $4N$. For the entire matrix we need $4N^2$ operations. Generally it is written as $OO(N^2)$ in an engineering sense, without much detail on the minutiae of the computations.

By performing either decimation in time or frequency once, as illustrated above, we have reduced the computations from $OO(N^2)$ to $OO(2(N/2)^2)$. The computational effort has been reduced by 50%. By repeating this till we get a two-point DFT, generating a DFT becomes generating an FFT and the effort will be $OO(N \log_2 N)$.

4.5 Fourier Series Coefficients

Fourier series are defined for periodic functions in time and also we assume that the integral of the function is finite. In this section we show the evaluation of Fourier series coefficients using a DFT. We take a pulse signal as an example.

4.5.1 Fourier Coefficients

Let $x(t)$ be a continuous and periodic function of t with periodicity T, defined as

$$x(t) = \begin{cases} A, & 0 \le t \le \tau, \\ 0, & \tau < t \le T, \end{cases} \qquad (4.26)$$

$$x(t) = \sum_{k=-\infty}^{\infty} \left\{ a_k \left(\cos\left[\left(\frac{2\pi k}{T} \right) t \right] \right) \right\} + \sum_{k=-\infty}^{\infty} \left\{ b_k \left(\sin\left[\left(\frac{2\pi k}{T} \right) t \right] \right) \right\}, \qquad (4.27)$$

where the values of a_k and b_k are defined as

$$a_k = \frac{2}{T} \int_{-T/2}^{T/2} x(t) \left[\cos\left(2\pi \frac{k}{T} t \right) \right] dt,$$

$$b_k = \frac{2}{T} \int_{-T/2}^{T/2} x(t) \left[\sin\left(2\pi \frac{k}{T} t \right) \right] dt. \qquad (4.28)$$

Let us consider a periodic function $x(t) = A$ in the interval $0 < t < \tau$, otherwise $x(t) = 0$. For this signal, the values of a_k and b_k are obtained by evaluating the integrals and are given as

$$a_k = \frac{A}{\pi k}\left(\sin\left[\frac{2\pi\tau}{T}k\right]\right) \quad \text{and} \quad b_k = \frac{2A}{\pi k}\left(\sin\left[\frac{\pi\tau}{T}k\right]\right)^2. \tag{4.29}$$

We have taken a numerical example with $\tau = 0.1$ s, $T = 1.2$ s and $A = 1.3$; we use a sampling time of 1 ms and a 512-point DFT. Figure 4.10 shows the theoretically computed coefficients superimposed with the coefficients computed using the DFT. There is a good match. Figure 4.10(a) shows the *even* nature of a_k and Figure 4.10(b) shows the *odd* nature of b_k. As a complex spectrum it is conjugate symmetric.

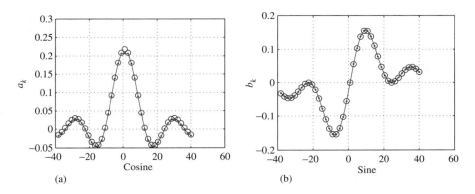

(a)   (b)

Figure 4.10   Computing Fourier coefficients using a DFT

4.6 Convolution by DFT

Convolution can be best understood in a *numerical* sense by looking at it as multiplication of two polynomials. We have taken this approach because polynomial multiplication is very elementary. Consider

$$\begin{aligned} p_1(x) &= 1 + 2x, \\ p_2(x) &= 1 + 3x + 4x^2. \end{aligned} \tag{4.30}$$

The product of the two polynomials $p(x) = p_1(x)p_2(x)$ is $1 + 5x + 10x^2 + 8x^3$ and the coefficients in $p(x)$ are the convolution of the two sequences $\{1, 2\}$ and $\{1, 3, 4\}$. A convolution operation involves many computations. Typically, if an m-sequence is to be convolved with an n-sequence, we need to perform $OO(mn)$ operations and the resulting sequence is of length $m + n - 1$. Taking the DFTs of the two sequences, multiplying them point by point and taking an inverse DFT

(IDFT) will give us the desired result. FFT offers a considerable advantage in performing convolution. Exercise caution as this will result in a circular convolution.[4] Let x_k and y_k be sequences of length m and n, respectively, then the circular convolution is defined as $x_k \odot y_k = \text{IDFT}\{(\text{DFT}(x_k) \times \text{DFT}(y_k)\}$.

4.6.1 Circular Convolution

In normal convolution the sequence beyond its length is assumed to have zeros. In circular convolution the sequences are assumed to be *periodic*, and this produces improper results if sufficient precautions are not taken. Consider a sequence x_k of length 50 ($m = 50$) and another sequence y_k of length 15 ($n = 15$) and we perform a normal convolution to get a sequence $z_k = x_k \otimes y_k$; this is the line marked with open circles in Figure 4.11(a) and has a length of 64 (50 + 15 − 1).

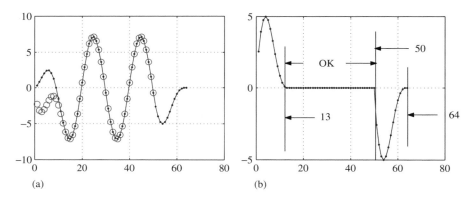

Figure 4.11   Understanding circular convolution

In Figure 4.11(a) two sequences were circularly convolved by multiplying the DFT of x_k with the DFT of y_k and taking the inverse DFT of the resulting sequence, leading to circular convolution $\hat{z}_k = x_k \odot y_k$. In order to take the DFTs and multiply them, we need the lengths to be the same. We have to make the lengths equal, so we have padded the sequence y_k with 35 zeros. The sequence $\hat{z}_k$ of length 50 is a consequence of this padding and is shown in Figure 4.11(a).

A comparison of the sequences z_k and $\hat{z}_k$ is shown as an error in Figure 4.11(b). It reveals that the first 12 points are erroneous and at the tail we need another 14 points of correct data. Figure 4.11(b) shows that only the length from 13 to 50 is numerically correct; the rest of the *head* and *tail* are not okay. To get over this problem we pad the sequence of length 50 with 14 ($= n - 1$) zeros and the sequence of length 15 with 49($= m - 1$) zeros, giving both sequences a length

---

[4]The symbol $\odot$ denotes circular convolution.

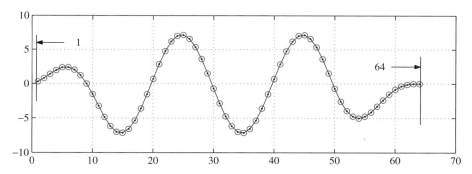

Figure 4.12   Circular convolution with right padding of zeros

of 64 ($= m + n - 1$). Using these two sequences, we perform circular convolution and the result is shown in Figure 4.12 along with the regularly convolved sequence.

4.7 DFT in Real Time

We have seen that a DFT can be computed by FFT or by recursive methods using a sliding DFT. Stability is a serious limitation of sliding DFTs. One of the most common requirements is averaging the spectrum to improve the SNR of the signal. According to Parseval's theorem, energy is conserved in the time domain or the frequency domain. And using this property, we realise that averaging in the time domain and computing the DFT is the same as computing the DFT and averaging it. Let the sequence of vectors $\mathbf{x}_1, \mathbf{x}_2, \ldots, \mathbf{x}_M$ have a corresponding sequence of power spectrum vectors $\mathbf{s}_1, \mathbf{s}_2, \ldots, \mathbf{s}_M$. Then

$$\frac{1}{M} \sum_{i=1}^{M} \mathbf{x}_i = \frac{1}{M} \sum_{i=1}^{M} \mathbf{s}_i \quad \text{where} \quad \mathbf{x}_i \xrightarrow{\|\text{DFT}\|} \mathbf{s}_i. \tag{4.31}$$

By performing a time average, we will miss the spectral update at regular intervals but we will gain on hardware complexity. What this means is that we need to wait until the $\sum_{i=1}^{M} \mathbf{x}_i$ computation has been completed before we can get the spectrum. To overcome this problem, inputs of two memory banks, A and B, are connected through a data switch S_1 (Figure 4.13). The switch S_1 is toggled between A and B when a bank is full. The output of the memory bank is also connected through a data switch S_2.

When the data is being filled in A, the FFT of the data in bank B is computed by positioning S_1 to A and S_2 to B. On each FFT computation we perform the recursive array averaging of (4.31). Therefore by pipelining we can get a constant spectral update.

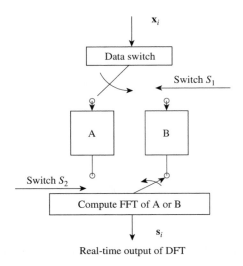

Figure 4.13   Real-time spectrum

4.7.1 Vehicle Classification

The idea of performing a time average is very important. We can use it for classifying automotive engines. Suppose we investigate a two-stroke engine and a four-stroke engine. A sequence of data z_k is collected as a .wav file using a conventional media player for a sufficiently long time, say 10 s. We choose a record $\mathbf{x}_i$ having size $m = 2^N$, part of z_k, such that when we concatenate all the $\mathbf{x}_i$ we will obtain the original sequence z_k. We average the vectors $\mathbf{x}_i$ and compute the DFT as

$$\mathbf{x}_\mu = \frac{1}{M} \sum_{i=1}^{M} \mathbf{x}_i \quad \text{and} \quad \mathbf{x}_\mu \xoverset{\|\text{DFT}\|}{\Longrightarrow} \mathbf{s}.$$

Here the block mean is replaced by the recursive mean using concepts in Section 1.7. Figure 4.14(a) shows the spectrum of the averaged vector $\mathbf{x}_\mu$ for a four-stroke engine and Figure 4.14(b) shows it for a two-stroke engine. Notice the distinct difference in spectral characteristics, allowing us to obtain a parametric signature. This averaging is essential in order to obtain the features of each spectrum.

4.7.2 Instrument Classification

This is a typical problem to upgrade the single micro-phone recording to a sterio recording. Let the time series recorded using a single micro-phone in a

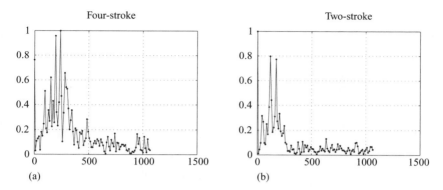

Figure 4.14   Time average spectrum

music concert be x_k. Then the problem formulation is to decompose the this time series as

$$x_k = z_k^1 + z_k^2 \ldots + z_k^N = \sum_{i=1}^{N} z_k^i$$

Each time series z_k^i corresponds to a specific instrument say *violin* as if they are recorded using a seperate micro-phone. This problem is very complex and is very important in replay proving a value to the listner.

4.8 Frequency Estimation via DFT

Signals are ordinarily *monochromatic* in nature, but discontinuities while acquiring a signal or intentional phase variations cause its *spectrum* to *spread*. Under these conditions it is difficult to estimate the frequency by conventional methods. Here we demonstrate the use of a DFT instead of an adaptive notch filter (ANF), even though there are other ways of tackling the problem. Problems of this type occur in underwater signal processing, radar signal processing and array signal processing. Making accurate frequency estimates is a significant problem. We consider an example from radar signal processing.

4.8.1 Problem Definition

Let each signal record r_i of length N be defined as

$$r_i(k) = \sum_{j=(i-1)N}^{j=iN-1} \sin(2\pi f j + \phi_i)\delta(k - j), \qquad (4.32)$$

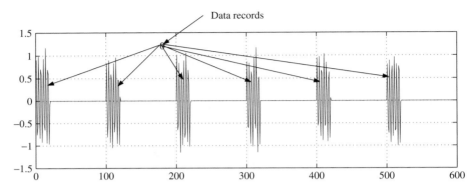

Figure 4.15 Pulses of data

where ϕ_i is a *uniformly distributed* random phase lying between π and $-\pi$ and $\delta(k - j)$ is defined as

$$\delta(k - j) = \begin{cases} 1 & \text{for} \quad k = j, \\ 0 & \text{otherwise.} \end{cases}$$

The problem is to find the frequency of the signal in the tone burst (Figure 4.15). Limitation comes from the fact that each record has a limited number of samples. So we create a new signal x_k obtained by assembling M such records and given as

$$x_k = \sum_{i=1}^{i=M} r_i(k). \tag{4.33}$$

Unfortunately, the acquired signal suffers from rounding of noise as well as unwanted signals. So the acquired signal y_k is modelled as

$$y_k = x_k + \gamma_k, \tag{4.34}$$

where γ_k is a white Gaussian noise (WGN) of zero mean and variance σ^2. Our current problem is to *accurately* estimate the quantity f. There are many methods available, but in this example we have chosen a DFT approach with curve fitting in the spectral domain.

4.8.2 DFT Solution

We demonstrate how to improve resolution with a limited number of samples by combining other techniques. We consider a time series as defined in (4.34) having $N = 20$ and $M = 6$ with a WGN of variance 0.5. We get a *total* record length of 120 samples.

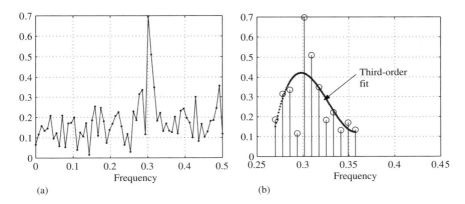

Figure 4.16 DFT of assembled records and curve fitting

For applications of this type it is best to take the DFT approach. Figure 4.16(a) shows the spectrum of the signal with a conspicuous maximum around 0.3 Hz. We have fitted a third-order polynomial around this maximum. For this polynomial we have obtained the maximum very accurately by computing the values for any desired frequency. This allows us to choose the resolution, leading to an accurate positional estimate[5] of the peak, as shown in Figure 4.16(b).

Often a simple DFT computation may not suffice on its own and we need to do more manipulation to obtain better results. In the case of heterodyning spectrum analysers, similar curve fitting is performed at the output of the bandpass filter.

4.9 Parametric Spectrum in RF Systems

Conventional methods cannot be applied for RF systems. Impulse response is irrelevant and these systems can only be characterised using s-parameters in the **S** matrix. However, the notion of a transfer function exists and we can make measurements of amplitude and phase.

Frequency domain least squares (FDLS) is useful for obtaining the transfer function of the RF device. In this approach, the system under consideration is excited by fixed frequencies and output amplitude and phase are measured at *steady state*. Suppose we obtain the following measurements using the experimental set-up in Figure 4.17. This set-up is typical of a network analyser. The staircase is used so the system can settle down and reach steady state for measurements.

Consider the unknown device to have a model given as

$$y_k = a_1 y_{k-1} + a_2 y_{k-2} + b_1 u_k + b_2 u_{k-1} + b_2 u_{k-2}. \tag{4.35}$$

---

[5]Location of the maximum rather than the value of maximum.

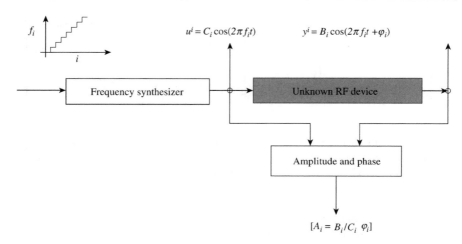

Figure 4.17   Frequency sweeping input for an unknown device

Let the input $u_k^i = \cos(2\pi\frac{fi}{N}k)$ and the output measured be $y_k^i = A^i\cos(2\pi\frac{fi}{N}k + \phi^i)$ under *steady-state* conditions. The term $\frac{fi}{N} = f_n$ is the normalised frequency. For each frequency f, measurements are made to obtain A^i and ϕ^i at *steady state*. For each measurement of A^i and ϕ^i at a specific frequency value of $\frac{fi}{N}$, y_k, y_{k-1}, y_{k-2}, u_k, u_{k-1}, and u_{k-2} are computed as follows:

$$y_k^i = A^i \cos\left[2\pi\frac{fi}{N}k + \phi^i\right], \tag{4.36}$$

$$y_{k-1}^i = A^i \cos\left[2\pi\frac{fi}{N}(k-1) + \phi^i\right], \tag{4.37}$$

$$y_{k-2}^i = A^i \cos\left[2\pi\frac{fi}{N}(k-2) + \phi^i\right], \tag{4.38}$$

$$u_k^i = \cos\left[2\pi\frac{fi}{N}k\right], \tag{4.39}$$

$$u_{k-1}^i = \cos\left[2\pi\frac{fi}{N}(k-1)\right], \tag{4.40}$$

$$u_{k-2}^i = \cos\left[2\pi\frac{fi}{N}(k-2)\right]. \tag{4.41}$$

Now we can rewrite (4.35) using the above data as

$$y_k^i = \left(\mathbf{g}^i\right)^t\mathbf{p}, \tag{4.42}$$

where $(\mathbf{g}^i)^t = [-y^i_{k-1}, -y^i_{k-2}, u^i_k, u^i_{k-1}, u^i_{k-2}]$ and $\mathbf{p}^t = [a_1, a_2, b_1, b_2, b_3]$. For brevity, we drop the subscript k and write (4.42) as

$$\mathbf{p}_N = \left(\sum_{i=1}^{N} \mathbf{g}^i (\mathbf{g}^i)^t \right)^{-1} \left(\sum_{i=1}^{N} \mathbf{g}^i y^i \right). \tag{4.43}$$

Steady-state measurement using a sweep frequency input signal offers several advantages over transient measurements such as the impulse response technique or time domain techniques [3]. Here are some of the advantages:

1. Less precise instrumentation is required for steady-state measurements than for transient measurements.
2. Collecting more data points improves the SNR for steady-state sweep frequency measurements, but not for transient measurements.
3. Steady-state measurements offer these advantages at the expense of longer measurement time.

4.9.0.1 Simulation Results

For the purpose of simulation, we consider a system with

$$B(z) = 0.126 + 0.252z^{-1} + 0.126z^{-2}, \tag{4.44}$$

$$A(z) = 1 - 0.89944z^{-1} + 0.404496z^{-2} \tag{4.45}$$

and we obtain output y_k as

$$y_k = \left(\frac{B(z)}{A(z)} \right) u_k + \gamma_k. \tag{4.46}$$

A moving-window DFT was used to monitor y_k until steady-state conditions were obtained. Then (4.43) was used to obtain the solution. Results are tabulated in Table 4.1. It is also possible to estimate the order of the system using this algorithm [4].

Table 4.1   Measurements obtained using the set-up in Figure 4.17

i	1	2	3	4	5	6	7	8	9	10
if_i/N	0.01	0.05	0.09	0.13	0.17	0.21	0.25	0.29	0.33	0.37
A_i	0.995	1.002	1.09	1	0.647	0.435	0.237	0.19	0.087	0.098
φ_i Deg	−4.73	−25.5	−47.3	−82.7	−113.5	−131.9	−141.2	−149.0	24.2	34.9

4.9.1 Test Data Generation

We require data to test the FDLS algorithm. This is generated synthetically. Consider a system

$$H(z) = \frac{B(z)}{A(z)}, \tag{4.47}$$

where $B(z)$ and $A(z)$ are chosen as in (4.44) and (4.45) and output y_k is obtained using (4.46). The *gain* A^i and the *phase* shift ϕ^i of the system at a specific frequency f, as required in (4.36) to (4.42), are obtained by choosing the frequency f^i then generating a sequence u_k as follows:

$$u_k = \cos(2\pi f_n k) \quad \text{where} \quad f_n = \frac{fi}{N} \text{ is the normalised frequency.} \tag{4.48}$$

The system is excited by u_k and for $k > 1000$ we obtain the DFT for the sequences u_k and y_k as

$$U = \text{DFT}(u_k w_k), \tag{4.49}$$

$$Y = \text{DFT}(y_k w_k), \tag{4.50}$$

where w_k is a time-domain weighting sequence known as a *Hamming window* and given by

$$w_k = \begin{cases} 0.54 + 0.46 \cos\left(\dfrac{\pi k}{N}\right), & |k| \leq N, \\ 0, & |k| > N. \end{cases} \tag{4.51}$$

Notice that the arrays U and Y are vectors with complex elements. Let $U = [U_1, \ldots, U_N]$. Since we know that the signal is a pure sinusoid, there will be a dominant peak in the spectrum. Hence the element of U whose magnitude is maximum is computed and the index where this maximum occurs is defined as j. Thus $j = \arg \max |U_l|$ for $1 \leq l \leq N$. Then we have

$$A^i_u = |U_j|, \tag{4.52}$$

$$\phi^i_u = \text{phase}(U_j). \tag{4.53}$$

Similarly, from the complex vector Y we can obtain A^i_y and ϕ^i_y. Now we generate the desired data as

$$A^i = \frac{A^i_y}{A^i_u}, \tag{4.54}$$

$$\phi^i = \phi^i_y - \phi^i_u. \tag{4.55}$$

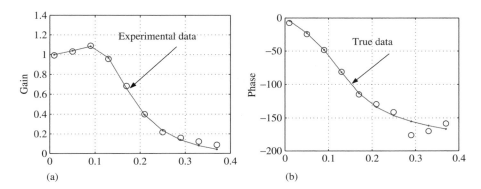

Figure 4.18   Frequency domain least-squares data

Figure 4.18 shows values of A^i and ϕ^i versus frequency at an SNR of 0 dB for the system defined by (4.44) and (4.45). True values are *superimposed* on the computed values. Each true value is obtained by sampling $H(z)$ on the unit circle at the desired frequency. Note that the values of A^i and ϕ^i are very accurate even when the SNR is 0 dB. This is the primary reason why the estimates are noise tolerant.

4.9.2 Estimating the Parameter Vector

Using the above data and (4.43), we get the solution as

$$\hat{\mathbf{p}} = (-0.9008 \quad 0.4055 \quad 0.1312 \quad 0.2502 \quad 0.1270).$$

The actual values can be obtained by looking at (4.44) and (4.45):

$$\mathbf{p} = (-0.8994 \quad 0.4045 \quad 0.1260 \quad 0.2520 \quad 0.1260).$$

4.10 Summary

In this chapter we used a matrix approach to explain the principles of the FFT. We have not used butterfly diagrams. Circular convolution was explained in depth. We presented a hardware scheme for implementation in real time. We looked at a problem of estimating frequency using a DFT. We elaborated a hardware structure for real-time implementation of continuous spectrum update. We ended by covering the principles of the network analyser, an example from RF systems.

References

1. E. O. Brigham, *The Fast Fourier Transform and Its Applications*. Englewood Cliffs NJ: Prentice Hall, 1988.
2. J. Cooley and J. Tukey, An algorithm for the machine calculation of complex Fourier series. *Mathematics of Computation*, **19**, 297–301, April 1965.
3. L. Ljung, *System Identification Theory for the User*, pp. 7–8. Englewood Cliffs NJ: Prentice Hall, 1987.
4. M. A. Soderstrand, K. V. Rangarao and G. Berchin, Enhanced FDLS algorithm for parameter-based system identification for automatic testing. In *Proceedings of the 33rd IEEE Midwest Symposium on Circuits and Systems*, Calgary, Canada, august 1990, pp. 96–99.

5

Realisation of Digital Filters

Translating difference equations into hardware is an important step in real life, and needs considerable care and attention. There are three distinct elements to this process (Figure 5.1). The first element is conversion of analogue signals to digital numbers; this is accomplished by an analogue-to-digital converter. The second element is implementation of the algorithm; this is carried out in the computing unit. The computing unit may be dedicated hardware made from adders, multipliers and accumulators or it may be a DSP processor. A dedicated unit can be an FPGA unit or a VLSI chip. The third element modifies the output of the computing unit in a manner required for further use in direct digital form or in analogue form. It varies from application to application. Sometimes we may require digitised data for processing.

5.1 Evolution

Hardware realisation of DSP algorithms, or for that matter all the embedded systems, has gone through a revolutionary change in recent years. The reason for this is the phenomenal growth rate of computing power. Clock speeds were around 1 MHz in 1976 whereas today they are 4 GHz or more. Clock speed is a good measure for estimating computing power.

From 1975 to 1985, development was more logic oriented and DSP was always on dedicated hardware. From about 1985 to 1995, it became more software oriented as hardware was becoming standardised while excellent software tools were being created to implement computation-intensive algorithms on single or multiple processors. From 1995 to the present, the line between hardware and software has become very narrow. For that matter, entire design is carried out on electronic design automation (EDA) tools using new methodologies like co-simulation, reducing the design cycle time and providing more time for testing, thus ensuring that the overall system performs closely to the desired specifications.

Digital Signal Processing: A Practitioner's Approach   K. V. Rangarao and R. K. Mallik
© 2005 John Wiley & Sons, Ltd

Figure 5.1   General block schematic for a DSP realisation

5.2 Development Process

The development process is depicted in Figure 5.2. There are nine distinct phases in the development process. These phases have remained the same throughout. Understanding the user application constitutes phase 1, based on which a specification document is generated in phase 2. This goes back and forth between the designer and the user, and finally a well-understood document is generated. In phase 3 a detailed design document is produced, which includes as many aspects of the problem as possible, including the development methodology. From 1975 to 1985, on completing phase 3 it was the general practice to go ahead with the coding and run the programs on the hardware by committing them to an EPROM or by testing them on local RAMs.

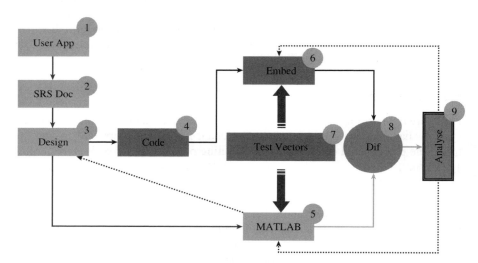

Figure 5.2   How to implement DSP algorithms

The test cycles (moving back and forth between phases 4 and 6) are extremely time-consuming; any errors in the design require an elaborate process of retracing.

Phases 3, 4, 6, 8 and 9 are quite often moved back and forth till the desired goal is achieved. From 1985 to 1995, another path entered the industrial process, one that used to be the domain of the universities. This path has a new phase, phase 5. It is a combination of phases 4 and 6, without hardware but close to reality. Most of the testing is done by moving back and forth in the path 3, 5, 8, 9. It occupies about 30% of the total development time.

Now, with computing power increasing exponentially, electronic design tools for VLSI development can be made available in *soft* form, so phases 4 and 6 have migrated into phase 5, leading to co-simulation.

5.3 Analogue-to-Digital Converters

In any hardware realisation for DSP, the most important component is the analogue-to-digital (A/D) converter 1. There are many commercial vendors specialising in many different approaches. Some of these approaches are described below.

5.3.1 Successive Approximation Method

Successive approximation is similar to the binary search that is practised in software. This scheme needs several important components: a very accurate digital-to-analogue (D/A) converter, analogue comparator, shift register and control logic. Let the signal be $y(t)$ then the D/A converter output is given as $\lceil \hat{y}_k \rceil$, where $\hat{y}_k$ is the digital input. The error signal is

$$e(t) = y_k - \lceil \hat{y}_k \rceil. \tag{5.1}$$

The DAC output is compared with the analogue signal. The binary signal $e(t)$ is the output of a high slew rate analogue comparator, and $\lfloor y_k \rfloor$ is the output of a shift register. Simple logic is used to shift and set the bit. If $e(t)$ is positive, the bit is set as 1 else 0 and a shift is made to the right. In Figure 5.3, S/H is a sample and hold and its output is y_k, SAR is a successive approximation register.

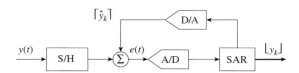

Figure 5.3 Successive approximation method

5.3.2 Flash Converters

In flash converters the given signal $x(t)$ is passed through a sample and hold to obtain x_k. This discretised value x_k is simultaneously compared with the weights of the full scale range and the result is produced as output of a comparator. Figure 5.4 shows a 3-bit flash A/D converter. The ith comparator output is given as

$$b_i = 0.5 \left[1 + \text{sign}\left(x_k - \frac{2^i - 1}{2^n} \right) \right], \tag{5.2}$$

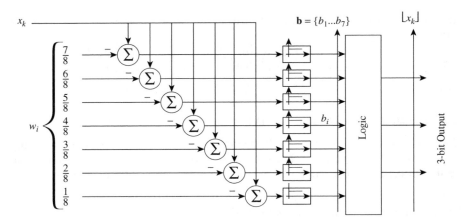

Figure 5.4   Flash analogue-to-digital converter

where n is the desired conversion width. Equation (5.2) is shown in Figure 5.4 implemented in analogue form. The weight $w_i = (2^i - 1)/2^n$ is obtained using a resistor in a potential divider formation. Comparators are high slew rate operational amplifiers specifically designed to meet the speed requirements. A vector $\mathbf{b} = [b_1, b_2, \cdots, b_{n-2}, b_{n-1}]$ is generated and encoded in the desired format using a logic function providing the digital output as $\lfloor x_k \rfloor$.

Notice this scheme has complex hardware. There are many methods available in the literature [4] to obtain higher-order conversions using 3-bit flash converters.

5.3.3 Sigma–Delta Converters

In sigma–delta converters (Figure 5.5) the signal $y(t)$ is compared with a binary signal b_k which indicates whether the signal $y(t)$ is in the top half or the bottom half of the conversion range:

$$e(t) = y(t) - b_k \quad \text{and} \quad z(t) = \int e(t)\, \mathrm{d}t \text{ is obtained.} \qquad (5.3)$$

The signal $z(t)$ is passed through a *hard limiter* generating a bitstream b_k.

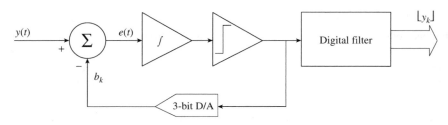

Figure 5.5   Sigma–delta converter

5.3.4 Synchro-to-Digital Converters

It may seem odd to talk about synchro-to-digital (S/D) converters but a close look indicates that they are functionally similar to A/D converters and offer an interesting mathematical viewpoint. The position on a rotating system is uniquely depicted as a 3-tuple $\Theta = [v_1 \ v_2 \ v_3]$. This vector is mapped into a scalar as $\Theta \rightarrow \theta$ through a table. In simple terms it is a mapping function from $\Re^3$ to $\Re$. This is mathematically the same as finding a direction using an antenna system (Section 6.2).

5.4 Second-Order BPF

Suppose the following transfer function is to be realised in hardware. The block schematic will be as shown in Figure 5.6.

$$H(z) = \frac{0.0062 - 0.0062z^{-2}}{1 - 1.608z^{-1} + 0.9875z^{-2}}. \tag{5.4}$$

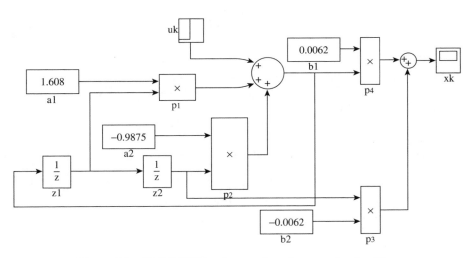

Figure 5.6 SIMULINK implementation of a second-order filter

The above z-domain filter can be readily rewritten as difference equations in the time domain:

$$w_k = 1.608w_{k-1} - 0.9875w_{k-2} + u_k, \tag{5.5}$$

$$x_k = 0.0062w_k - 0.0062w_{k-2}. \tag{5.6}$$

This requires us to perform four multiplications and three additions. We can see that not all the multiplications and additions are serial. The way we have written the original filter (5.4) as a set of two independent difference equations with minimum delays gives an important clue about the realisation. Equation (5.5) needs two multiplications and two additions, and the multiplications can be carried out simultaneously. Figure 5.6 shows a SIMULINK[1] implementation of the filter. There are many third-party vendors providing code generation packages on your chosen processors.

5.4.1 Fixed-Point Implementation

We have directly implemented a filter given as $H(z)$ into floating-point hardware. This may not always be the case. For greater speed and for other reasons such as cost and maintainability, the fixed-point implementation is still a very useful architecture.

This needs a different approach. As a first step, we need to convert the filter coefficients into rational numbers (of the form p/q). The denominator will be a number of the form 2^N, where N is an integer:

$$w_k = \left(\frac{411}{256}\right) w_{k-1} - \left(\frac{252}{256}\right) w_{k-2} + \left(\frac{2^{16}}{256}\right) u_k,$$

$$x_k = 256 w_k - 256 w_{k-2}. \tag{5.7}$$

In this numerical example we have $N = 8$. This means the input is first required to be shifted left 16 places for making a multiplication with 2^{16}. It means that we need a 24-bit adder if the input is an 8-bit A/D converter. On making fixed-point multiplication and addition, we shift right 8 times for dividing by 256 to generate w_k. The fixed-point implementation is shown in Figure 5.7. Each register provides the delay and the registers are clocked with a *minimum* period (t_c) equal to the time taken for one multiplication (t_*) and one addition (t_+), so $t_c \geq t_* + t_+$. It means that the output of the register settles only after period $t_d = t_s + t_* + t_+$ where t_s is the register *set-up* time. We have used two registers r_1 (input r_1^i and output r_1^o) and r_2 in the above hardware and the information is sequenced as follows (Figure 5.7).

$$r_1^i \Leftarrow [r_2^o + (2^{16})u]2^{-8} \qquad \text{Settling time} = t_+, \tag{5.8}$$

$$r_2^i \Leftarrow -252r_1^o + 411r_1^i \qquad \text{Settling time} = t_* + t_+ + \delta t, \tag{5.9}$$

$$r_1^i \longrightarrow r_1^o \; r_2^i \longrightarrow r_2^o \qquad \text{Clocking of registers.}$$

Clocking of the registers can be done *only after* the combinatorial circuit outputs connected to the inputs of the two registers (r_1^i and r_2^i) settle. Figure 5.8 compares the fixed-point filter with the floating-point filter.

---

[1]A software package from MathWorks of the US.

$$w_k = \left(\frac{411}{256}\right) w_{k-1} - \left(\frac{252}{256}\right) w_{k-2} + \left(\frac{2^{16}}{256}\right) u_k$$

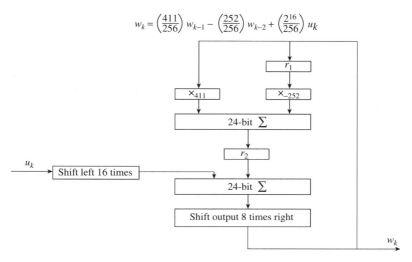

Figure 5.7 Fixed-point IIR filter

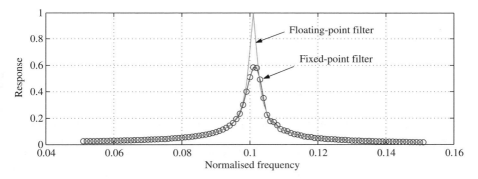

Figure 5.8 Comparison of floating-point and fixed-point filters

5.5 Pipelining Filters

Often the throughput rate is more important than the latency in filter designs. Pipelining is the art of positioning the hardware blocks such that the throughput rate is maximised. Output results are available after every clock cycle.

5.5.1 Modelling Hardware Multipliers and Adders

We introduce $\hat{+}$ representing a hardware adder and $\hat{*}$ representing a multiplier. In reality, all arithmetic operations take a finite amount of time, introducing a delay into the system. Multiplication of two variables, say b_1 and u_k, in hardware can be modelled as

$$b_1 \hat{*} u_k = b_1 * u_k D_1 = b_1 u_k D_1, \tag{5.10}$$

where $*$ is the *ideal* multiplier and D_1 is the propagation delay associated with the multiplication. The *ideal* adder is given as $+$ and the corresponding delay associated with the adder is D_2. Then an addition of two variables u_k and u_{k-1} can be written as

$$u_k \hat{+} u_{k-1} = (u_k + u_{k-1}) D_2. \tag{5.11}$$

Each of the previous devices can be incorporated into a pipelined digital filter structure by following it with a synchronising register. The clock period must then be less than $D + t_s + t_r$, where t_s is the register set-up time and t_r is the register propagation delay. With this notation we can rewrite (5.5) as

$$w_k = 1.608 \hat{*} w_{k-1} \hat{-} 0.9875 \hat{*} w_{k-2} \hat{+} u_k \tag{5.12}$$

$$= 1.608 * w_{k-1} D^{-1} - D^{-1} 0.9875 * w_{k-2} D^{-1} + D^{-1} u_k. \tag{5.13}$$

5.5.2 Pipelining FIR Filters

In this section we wish to realise

$$x_k = b_1 * u_k + b_2 * u_{k-1} + b_3 * u_{k-2}. \tag{5.14}$$

It is acceptable to have an overall delay D^N but it is desirable to minimise N:

$$\hat{x}_k = D^N \{ b_1 * u_k + b_2 * u_{k-1} + b_3 * u_{k-2} \}. \tag{5.15}$$

First, by associating one D with each multiplier and forming real multipliers as per the notation $D* = \hat{*}$, we get

$$\hat{x}_k = D^{N-1} \{ b_1 D * u_k + b_2 D * u_{k-1} + b_3 D * u_{k-2} \}. \tag{5.16}$$

Now the real multiplier is formed:

$$\hat{x}_k = D^{N-1} \{ b_1 \hat{*} u_k + b_2 \hat{*} u_{k-1} + b_3 \hat{*} u_{k-2} \}. \tag{5.17}$$

Next we use one additional D to associate with the adding of $b_2 \hat{*} u_{k-1}$ and $b_3 \hat{*} u_{k-2}$. It is important to choose the most delayed values first in order to minimise N.

$$\hat{x}_k = D^{N-2} \{ Db_1 \hat{*} u_k + D(b_2 \hat{*} u_{k-1} + b_3 \hat{*} u_{k-2}) \}. \tag{5.18}$$

Now, forming the real adder,

$$\hat{x}_k = D^{N-2}\{Db_1 \,\hat{*}\, u_k + (b_2 \,\hat{*}\, u_{k-1}\,\hat{+}\, b_3 \,\hat{*}\, u_{k-2})\}, \tag{5.19}$$

and using an additional delay D to associate with the final adder, a real adder is obtained:

$$\hat{x}_k = D^{N-3}\{Db_1 \,\hat{*}\, u_k \,\hat{+}\, (b_2 \,\hat{*}\, u_{k-1} \,\hat{+}\, b_3 \,\hat{*}\, u_{k-2})\} \tag{5.20}$$

Finally, we choose the system delay such that $D = z^{-1}$ and replace $u_{k-2} = z^{-2}u_k$ and $u_{k-1} = z^{-1}u_k$, so we get

$$\hat{x}_k = z^{-N+3}\{z^{-1}b_1 \,\hat{*}\, u_k \,\hat{+}\, (z^{-1}b_2 \,\hat{*}\, u_k \,\hat{+}\, z^{-2}b_3 \,\hat{*}\, u_k)\}. \tag{5.21}$$

Removing the factor z^{-1} from the expression, we obtain

$$\hat{x}_k = z^{-N+2}\{b_1 \,\hat{*}\, u_k \,\hat{+}\, (b_2 \,\hat{*}\, u_k \,\hat{+}\, z^{-1}b_3 \,\hat{*}\, u_k)\}. \tag{5.22}$$

Note that $N = 2$ is the minimum possible value of N for a causal filter. Choosing $N = 2$ gives

$$\hat{x}_k = \{b_1 \,\hat{*}\, u_k \,\hat{+}\, (b_2 \,\hat{*}\, u_k \,\hat{+}\, z^{-1}b_3 \,\hat{*}\, u_k)\}. \tag{5.23}$$

Equation (5.23) is very convenient for pipelining and is illustrated in Figure 5.9. Hardware enclosed by a dotted line represents a multiplier ($\hat{*}$) or an adder ($\hat{+}$), taking propagation delays into account. The above procedure is general for any FIR filter. Pipelining an FIR filter results in maximising the sampling or throughput frequency at the expense of a delay or latency in the availability of the output. It is like reading yesterday's paper or the day before yesterday's, every morning.

5.5.3 Pipelining IIR Filters

Pipelining an IIR filter is considerably more difficult than pipelining an FIR filter. This is mainly because the delay in the availability of the output makes it impossible to feed back output values with short delay, as is required for the straightforward IIR implementation. As an example, consider a simple first-order IIR filter

$$x_k = a_1 * x_{k-1} + u_k. \tag{5.24}$$

As in the FIR case, we introduce a delay D^N as follows:

$$\hat{x}_k = D^N(a_1 * x_{k-1} + u_k). \tag{5.25}$$

$$\hat{x}_k = \{b_1 \,\hat{*}\, u_k \,\hat{+}\, (b_2 \,\hat{*}\, u_k \,\hat{+}\, z^{-1}b_3 \,\hat{*}\, u_k)\} \qquad (5.23)$$

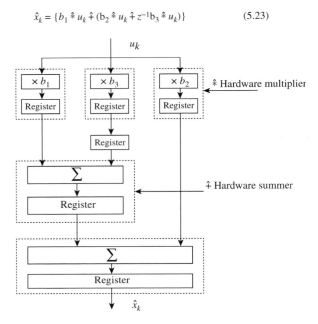

Figure 5.9 Pipelining of an FIR filter

Now we use one delay to form the real multiplier:

$$\hat{x}_k = D^{N-1}(a_1 D * x_{k-1} + Du_k) \qquad (5.26)$$
$$= D^{N-1}(a_1 \,\hat{*}\, x_{k-1} + Du_k). \qquad (5.27)$$

We use one more D to form the real adder

$$\hat{x}_k = D^{N-2}(a_1 D * x_{k-1} \,\hat{+}\, u_k). \qquad (5.28)$$

Finally, we put $D = z^{-1}$ and $x_{k-1} = z^{-1}x_k$ to obtain

$$\hat{x}_k = z^{-N+2}(a_1 \,\hat{*}\, z^{-1}x_k \,\hat{+}\, z^{-1}u_k) \qquad (5.29)$$
$$= z^{-N+1}(a_1 \,\hat{*}\, x_k \,\hat{+}\, u_k). \qquad (5.30)$$

For $\hat{x}_k = x_k$, N must be zero. However, $N \geq 1$ is required for a causal filter.

Solutions to this problem exist. They use a higher-order difference equation representation of the filter with equivalent characteristics, so feedback loop delay greater than 1 can be tolerated. More details can be found in the literature [2] on pipelining IIR digital filters. In order to pipeline an IIR filter, consider a filter

$$x_k = a_3 x_{k-3} + a_4 x_{k-4} + u_{k-1}. \tag{5.31}$$

The IIR filter defined in (5.31) can be represented using the hardware multipliers as

$$x_k = (x_k \,\hat{*}\, a_3 \,\hat{+}\, z^{-1} x_k \,\hat{*}\, a_4) \,\hat{+}\, u_k. \tag{5.32}$$

Figure 5.10 shows the hardware equivalent of (5.32); adders ($\hat{+}$) and multipliers ($\hat{*}$) are enclosed by dotted lines. This is to indicate that certain IIR filters of this class can be pipelined. Taking this as a hint, we proceed as follows.

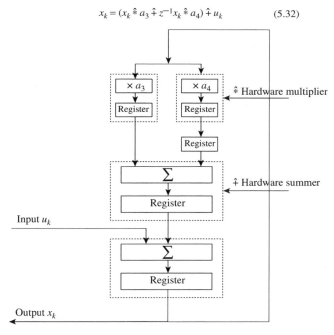

Figure 5.10   Pipelining of an IIR filter

Consider a second-order filter

$$H(z) = \frac{1}{A(z)} \quad \text{where} \quad A(z) = 1 - rpz^{-1} + r^2 z^{-2}, \tag{5.33}$$

in which $p = 2 \cos \theta$ is required to be pipelined. We have demonstrated that all FIR filters can be pipelined and only some IIR filters can be pipelined. With this backdrop, it is proposed to modify the filter in (5.33) as

$$H(z) = \left[\frac{B(z)}{A(z)}\right]\frac{1}{B(z)}$$

$$= B(z)\left[\frac{1}{A(z)B(z)}\right]. \tag{5.34}$$

The choice of $B(z)$ is such that the product $D(z) = B(z)A(z)$ is of the form shown in (5.31) and (5.32):

$$B(z)A(z) = 1 - a_3z^{-3} - a_4z^{-4}$$

$$= 1 + r^3(2p - p^3)z^{-3} + r^4(p^2 - 1)z^{-4}. \tag{5.35}$$

A filter of the form $D(z)$ is highly pipelinable. The choice of $B(z)$ is given as

$$B(z) = 1 + rpz^{-1} + r^2(p^2 - 1)z^{-2}. \tag{5.36}$$

We can see that $H(z)$ is composed of $B(z)$, an FIR filter which is pipelinable, and $1/D(z)$ and is of the form given by the difference equation (5.31), which is also pipelinable. This implies that $H(z)$ can be pipelined, modifying the structure as $H(z) = B(z)/D(z)$. This method is good but caution must be exercised over the stability aspects.

5.5.4 Stability Issues

We need to look at the stability of the new filter $H(z) = B(z)/D(z)$. The denominator polynomial is $D(z) = B(z)A(z)$, in which the polynomial $A(z)$ is stable. Hence we need to examine the roots of the polynomial $B(z)$ (5.36). Figure 5.11 shows the roots of $B(z)$ at various values of θ for $r = 0.9$. The *unstable*

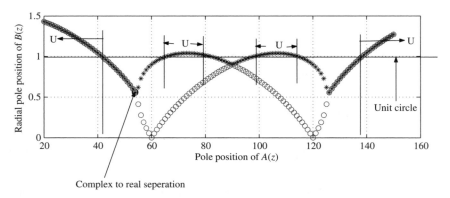

Figure 5.11   Stability of polynomial $D(z)$ in pipelining

regions are marked U. This imposes a limitation on using the entire spectrum, since roughly 50% of the sampling zones are unstable. If we compromise on the the sharpness of the filter, then these zones can be reduced. But by compromising on sharpness, a higher-order FIR filter can probably be used instead of this structure.

5.6 Real-Time Applications

In real-time applications, a single filter or set of filters operate on the data and have to provide the results in a specified time demanded by the overall system requirements. Depending on the time criticality, real-time applications can be implemented using a single processor or using a combination of processor and dedicated hardware. The best examples are the FFT chipsets that were very popular from 1985 to 1993. There are many programable digital filter chips on the market and they can be interfaced to another processor, which may be of low computing power, as a trade-off between cost and hardware complexity. With the availability of powerful processors, almost all the functions can be directly implemented on the same processor.

5.6.1 Comparison of DSP Processors

Table 5.1 has been compiled from published literature issued by commercial vendors of DSP processors.[2] One important performance measure is the clock frequency or million instructions per second (MIPS). At periodic intervals, Berkeley Design Technolgy Inc. (BDTI) publishes its own processor evaluations based on benchmark programs it has designed.

Table 5.1 Benchmarking some commercial DSP processors

	Feature	
Processor	Frequency or MIPS	BDTI mark
ADI ADSP-2183	52	13
ARM7TDMI/Piccolo	70	14
Hitachi SH-DSP	66	17
Intel Pentium	233	26
Lucent DSP1620	120	22
Lucent DSP16210	100	36
Motorola DSP56011	47.5	13

---

[2]The data is not from a single source but from various DSP processor vendors; it is only informative in nature.

5.7 Frequency Estimator on the DSP5630X

A DSP algorithm implementation on a processor in real time involves three distinct steps:

1. Rapid prototyping using MATLAB or any other higher-level language
2. Translation to ANSI C and testing the same
3. Using a cross compiler to download the code on to the DSP processor

We have implemented this methodology (Figure 5.12) on a DSP5630X processor. Approximately 40% of total development time is invested on algorithm development and 30% is for testing the algorithm.

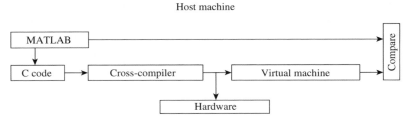

Figure 5.12   Algorithm to hardware

5.7.1 Modified Response Error Method

In the modified response error method [3], the signal y_k is modelled as an output of an oscillator corrupted by a narrowband noise ψ_k and WGN γ_k. So the modelled signal $\hat{y}_k$ can be written as

$$\hat{y}_k = [\sin(2\pi f k) + \psi_k] + \gamma_k. \tag{5.37}$$

The quantity $\sin(2\pi f k)$ in (5.37) can be modelled as an output of an autoregressive second-order system, AR(2), with poles on the unit circle, and is given by

$$\hat{x}_k = \hat{a}_1 \hat{x}_{k-1} + \hat{a}_2 \hat{x}_{k-2}, \tag{5.38}$$

where $\hat{a}_1 = 2\cos(\omega)$ and $\hat{a}_2 = -1$. However, in general $a_1 = 2r\cos(\beta)$ and $a_2 = -r^2$ where $re^{\pm j\beta}$ are the complex conjugate poles. A distinct peak exists [8] at ω_p, which is given by

$$\omega_p = \cos^{-1}\left[0.25a_1\left(1 + \frac{1}{a_2}\right)\right]. \tag{5.39}$$

In this problem we shall simultaneously estimate the parameters and the state of the system.

5.7.1.1 Parameter Estimation

In parameter estimation we consider the error $e_k = y_k - \hat{x}_k$ and the objective function $J_k = e_k^2 - v$, where the parameter v is also estimated along with $\hat{a}_1$ and $\hat{a}_2$. The choice of v is such that $E[J_k] = 0$. For minimising the objective function J_k, gradients are obtained by differentiating (5.38) with respect to $\hat{a}_1$ and we get

$$s_k = -\frac{\partial e_k}{\partial \hat{a}_1}$$

$$= \frac{\partial \hat{x}_k}{\partial \hat{a}_1} = \hat{a}_1 \frac{\partial \hat{x}_{k-1}}{\partial \hat{a}_1} + \hat{a}_2 \frac{\partial \hat{x}_{k-2}}{\partial \hat{a}_1} + \hat{x}_{k-1}. \tag{5.40}$$

Using (5.40) we obtain the gradient vector as

$$\left(\frac{\partial J_k}{\partial \mathbf{p}_k}\right)^t = [2e_k s_{k-1}, \ 2e_k s_{k-2}, \ 1]$$

$$= [\nabla J(\mathbf{p}_k)]^t$$

$$= \mathbf{g}_k^t, \tag{5.41}$$

where parameter $\mathbf{p}_k$ is given as

$$\mathbf{p}_k^t = [\hat{a}_1, \ \hat{a}_2, \ \hat{v}] \tag{5.42}$$

and the value s_k is obtained by recasting (5.40):

$$s_k = \hat{a}_1 s_{k-1} + \hat{a}_2 s_{k-2} + \hat{x}_k. \tag{5.43}$$

We can approximate (5.43) as

$$s_k = \hat{a}_1 s_{k-1} + \hat{a}_2 s_{k-2} + y_k. \tag{5.44}$$

This is due to the fact that (5.43) represents a *high-Q filter* very close to convergence, hence we can drive the filter using y_k to obtain s_k due to non-availability of $\hat{x}_k$.

5.7.1.2 Parameter Incrementation

The general incrementation procedure [2] for minimising the function $J(\mathbf{p}_k)$ is given as

$$\mathbf{p}_{k+1} = \mathbf{p}_k + \left[\nabla^2 J(\mathbf{p}_k)\right]^{-1} [\nabla J(\mathbf{p}_k)] \mu_k. \tag{5.45}$$

In the current problem we choose the Hessian [6] as

$$[\nabla^2 J(\mathbf{p}_k)] = \left[\sum_{i=1}^{k} \mathbf{g}_i \mathbf{g}_i^t \right]$$
$$= \mathbf{H}_k^{-1}, \tag{5.46}$$

and the parameter is incremented [7] via

$$\delta \mathbf{p}_k = \mathbf{H}_k \mathbf{g}_k \mu_k, \tag{5.47}$$
$$\mathbf{p}_{k+1} = \mathbf{p}_k + \delta \mathbf{p}_k. \tag{5.48}$$

A *heuristic* choice of the step size μ_k was chosen as the predicted objective function given by

$$\mu_k = \hat{J}_k = J_k - \mathbf{g}_k \delta \mathbf{p}_{k-1}^t. \tag{5.49}$$

5.7.1.3 State Adjustment

In state adjustment, the AR(2) process (5.38) which serves as a model has no input added to that at convergence; this AR(2) process is an *oscillator*. It makes parameter incrementation unstable under normal circumstances. It was found that state adjustment has a *stabilising* effect and is given by

$$\delta \hat{x}_k = s_{k-1} \delta \hat{a}_1 + s_{k-2} \delta \hat{a}_2. \tag{5.50}$$

5.7.1.4 Frequency Estimation

Frequency is computed by estimating the parameters $\hat{a}_1$ and $\hat{a}_2$ and substituting them in (5.39). Note that β is different from ω_p.

5.7.1.5 Recursive Estimation

The above algorithm is conveniently represented recursively by using the matrix inversion lemma [1] for obtaining $\mathbf{H}_k$. This algorithm works from arbitrary initial conditions by using a *forgetting factor* [5] α_k while computing $\mathbf{H}_k$. The multi-resolution regularised expectation maximisation (MREM) algorithm is as follows:

$$\mathbf{p}_k^t = [\hat{a}_1, \ \hat{a}_2, \ v], \tag{5.51}$$
$$\hat{x}_k = \hat{a}_1 \hat{x}_{k-1} + \hat{a}_2 \hat{x}_{k-2}, \tag{5.52}$$
$$s_k = \hat{a}_1 s_{k-1} + \hat{a}_2 s_{k-2} + y_k, \tag{5.53}$$
$$e_k = y_k - \hat{x}_k, \tag{5.54}$$
$$J_k = e_k^2 - v, \tag{5.55}$$
$$\mathbf{g}_k^t = [2e_k s_{k-1}, \ 2e_k s_{k-2}, \ 1], \tag{5.56}$$

$$\mathbf{H}_k = (\mathbf{H}_{k-1}/\alpha_k) + \left[\frac{(\mathbf{H}_{k-1}/\alpha_k)\mathbf{g}_k\mathbf{g}_k^t(\mathbf{H}_{k-1}/\alpha_k)}{1 + \mathbf{g}_k^t(\mathbf{H}_{k-1}/\alpha_k)\mathbf{g}_k}\right], \quad (5.57)$$

$$\hat{J}_k = J_k - \mathbf{g}_k\delta\mathbf{p}_{k-1}^t, \quad (5.58)$$

$$\delta\mathbf{p}_k = \mathbf{H}_k\mathbf{g}_k\hat{J}_k$$

$$= [\delta\hat{a}_1, \ \delta\hat{a}_2, \ \delta v]^t, \quad (5.59)$$

$$\mathbf{p}_{k+1} = \mathbf{p}_k + \delta\mathbf{p}_k, \quad (5.60)$$

$$\delta\hat{x}_k = s_{k-1}\delta\hat{a}_1 + s_{k-2}\delta\hat{a}_2, \quad (5.61)$$

$$\hat{x}_k = \hat{x}_k + \delta\hat{x}_k, \quad (5.62)$$

$$\alpha_k = \alpha_{k-1}0.9995 + 0.0005. \quad (5.63)$$

5.7.2 Algorithm to Code

The above algorithm was first implemented in MATLAB and then a C code was written for easy implementation on the Motorola DSP5630X. Both the MATLAB code and the corresponding C code are given in the appendix. Figure 5.13(a) shows the input and the estimated output using the MREM algorithm presented as above. In Figure 5.13(b) we can see the convergence of the parameters. A quick observation shows that a_2 is almost one whereas a_1 is representative of the frequency.

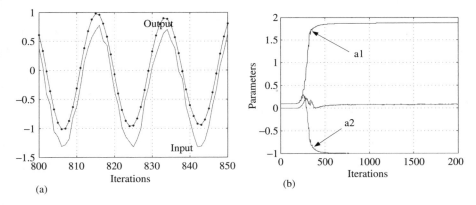

Figure 5.13   Implementation on a DSP5630X

5.8 FPGA Implementation of a Kalman Filter

In this section we implement the algorithm developed in Section 3.11.1. The algorithm is focused on reducing complexity to two FIR filters. Algorithm development is the most critical part of any implementation. In this case the critical aspect is reducing the complex calculations to FIR filters for ease of implementation on field-programable gate arrays (FPGAs). FPGAs are combinatorial circuits which map the design in the form of lookup tables (LUTs). Here are the steps we followed once we had developed the algorithm:

1. Algorithm prototype using MATLAB: the rate estimator algorithm was designed using MATLAB and tested.
2. Verilog code for FPGA: the *derivator* was coded in Verilog, simulated on ACTIVE-HDL and synthesised on Leonardo Spectrum. For representing the real numbers we used a sign magnitude fixed-point implementation.

5.8.1 Fixed-Point Implementation

The design of the Kalman filter consists of two FIR filters. FPGAs are better suited to implementing FIR filters than IIR filters. The inputs to the Kalman filter are real numbers; they are represented in fixed-point format rather than floating-point format as floating-point adders and multipliers occupy more chip area on an FPGA. The fixed-point representation is taken to be 6 bits wide since the area occupied on the FPGA will be less. Of these 6 bits, the most significant bit (MSB) is used as a sign bit (zero indicates a positive number and one indicates a negative number), 4 bits are used for the fractional part and 1 bit is used for the integer part (since we are considering normalised data input).

The reason for choosing a sign magnitude representation is the ease of implementation, even though twos complement has a much higher accuracy. The code flow includes implementation of sign magnitude fixed-point adders and fixed-point multipliers.

The implementation of the rate estimator was coded in Verilog, simulated on ACTIVE-HDL and synthesised on Leonardo Spectrum. The comparative results obtained from ACTIVE-HDL and MATLAB are shown in Figure 5.14(a) and (b). The target device was the VIRTEX-II PRO. The FPGA results are clipped at the

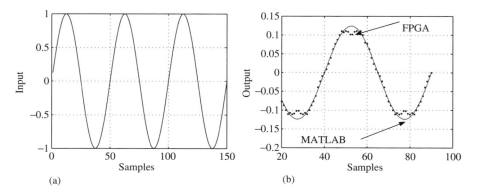

Figure 5.14   Implementation on an FPGA

Table 5.2 Synthesis report for the FPGA implementation

Resources	Used	Avail	Utilisation
Input/output	32	140	22.86
Global buffers	0	16	0.00
Function generators	451	2816	16.02
CLB slices	226	1408	16.05
DFFs or latches	0	3236	0.00
Block RAMs	0	12	0.00
Block multiplier	4	12	33.33
Total FPGA	**713**	**7640**	**9.33**

peak (Figure 5.14(b)), but this is only due to the lower resolution (4 bits). The synthesis report in Table 5.2 gives the silicon utilisation. CLBs are configurable logic blocks and DFFs are delay flip-flops.

5.9 Summary

In this chapter we described various hardware elements that are frequently used to convert the given analogue signal into digital form. We implemented a second-order BPF in SIMULINK and in fixed-point hardware then compared the results. We looked at a hardware realisation scheme that led us to consider pipelining and its associated problems. We gave an in-depth discussion on various methods of pipelining FIR and IIR filters. Then we looked at a complex algorithm for estimating frequency based on time series data and implemented on the DSP5630X using C. To implement this algorithm using FPGAs, we chose a design using FIR filters.

References

1. K. H. Lundberg, High-speed analogue-to-digital converter survey. Available from www.mit.edu/klund/www/papers/UNP_flash.pdf.
2. M. A. Soderstrand, H. H. Loomis and R. Gnanasekaran, Pipelining techniques for IIR digital filters. In *Proceedings of the IEEE International Symposium on Circuits and Systems*, New Orleans LA, May 1990, pp. 121–24.
3. K. V. Rangarao and P. A. Janakiraman, A new algorithm for frequency estimation in the presence of phase noise. *Digital Signal Processing*, **4**(3), 198–203, July 1994.
4. M. Yotsyanagi, T. Etoh and K. Hirata, A 10-b 50-MHz pipelined CMOS A/D converter with S/H. *IEEE Journal of Solid State Circuits*, **28**(3), 292–300, March 1993.
5. T. Söderström and P. Stoica, *System Identification*. Hemel Hempstead, UK: Prentice Hall, 1989.
6. P. Young, *Recursive Estimation and Time Series Analysis*, p. 25. New York: Springer-Verlag, 1984.
7. D. M. Himmelblau, *Applied Non-linear Programming*, p. 75. New York: McGraw-Hill, 1972.
8. M. S. El Hennawey *et al.*, MEM and ARMA estimators of signal carrier frequency. *IEEE Transactions on Acoustics, Speech and Signal Processing*, **34**, 618, June 1986.

6

Case Studies

Digital signal processing is all-pervading, so we have mostly selected the areas that interest engineers. Often the solutions look very similar, indicating many common-alities between them. These case studies also demonstrate that it is very difficult to categorise a given problem as exclusively DSP or non-DSP. Frequently we need to stretch our *engineering* hand a little and remove our *blinkers* to get an overall view of the solution.

In addition, problems are non-linear and we need a good systems background as well as DSP. Inverse problems are more difficult than forward problems or synthesis. For instance, the design of a filter for a given specification is trivial. It is the combinations of many disciplines put together that makes a problem very interesting. Perhaps you feel that problems are centred around some common solutions. But this chapter aims to demonstrate that the essence of any solution is a good problem formulation.

6.1 Difference Equation to Program

The following C code is for a digital filter implementation that includes testing by exciting it with a white Gaussian noise (WGN); the output is captured in the file `filter.dat`. The spectral characteristic of the output time series is the same as that of the filter, since the input is a WGN. This code demonstrates only that converting a given difference equation to a code is relatively trivial. Here we designate $xk1 = x_k$, $xk2 = x_{k-1}$, $xk3 = x_{k-2}$ and the filter is $x_k = 0.5871\,x_{k-1} - 0.9025 * x_{k-2} + 0.0488 * (u_k - u_{k-2})$.

```
// Translating a difference equation to C

# include <stdio.h>
# include <stdlib.h>
```

Digital Signal Processing: A Practitioner's Approach K. V. Rangarao and R. K. Mallik
© 2005 John Wiley & Sons, Ltd

```
# include <math.h>

float randn(void); FILE * fp_write;
void main (void)
{
int k, kmax=5000;
float pi, yk, f=0.1; float xk1=0.0, xk2=0.0, xk3=0.0;
float uk1=0.0, uk2=0.0, uk3=0.0;
fp_write = fopen("filter.dat", "w");
pi = (atan(1.0))*4.0; randomise();
printf(" start \n");
         for (k=0; k < kmax; k++)
              {
                  uk1=randn(); xk3=xk2; xk2=xk1;
                  uk3=uk2; uk2=uk1;
         xk1=0.5871*xk2 - 0.9025*xk3 + 0.0488*(uk1-uk3);
                  fprintf(fp_write,"%d %f \n", k, xk1);
              }
printf(" finish \n");
fclose(fp_write);
}
```

The following program generates white Gaussian noise by adding 12 uniformly distributed random variables.

```
float randn(void)
{
float intmax, udf, sum;
int rnd, k;

intmax = RAND_MAX;
      sum = 0.0;
      for (k = 0; k < 12; k++)
      {
         rnd = rand(); udf = rnd;
         udf = (udf/intmax)-0.5; sum = sum + udf;
      }
return(sum);
}
```

6.2 Estimating Direction of Arrival

Estimating direction of arrival (DoA) is of continual interest to people working on underwater systems or RF systems. Locating a radiating source [1, 6, 7] requires us to determine the DoA of energy from a mobile phone, a push-to-talk or an underwater acoustic emitter. Monopulse radar [5] also works on the same principle, as shown below.

To understand this in simple terms, let us consider three directional receivers as shown in Figure 6.1. Each directional receiver consists of a set of antenna elements and a receiver that has a definite gain pattern. The first receiver provides an output r_1 in a given direction θ. To be more precise, we assume an emitter of a given power when moved at a constant distance. Put another way, in a circle around this directional receiver, the output r_1 follows the pattern in Figure 6.1.

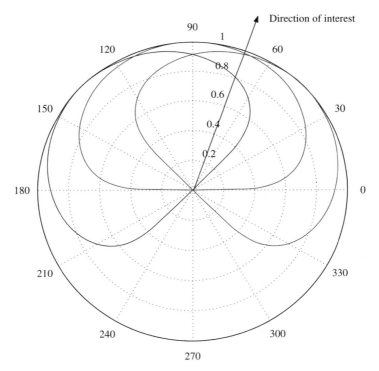

Figure 6.1 Directional receivers $r = |\sin\theta|^{0.25}$ for $\theta = 0$ to π

When we have three such receivers each shifted by $45°$ we get a response vector $\Theta_i = [r_1^i \ r_2^i \ r_3^i]$, where each element $(r_{1,2,3}^i)$ represents the response of each receiver. Now we map this 3-tuple Θ to a scalar θ. Let the ensemble of vectors be $\mathbf{A} = [\Theta_1 \ \Theta_2 \ldots \Theta_N]$. We can see that $\mathbf{A}$ is an $N \times 3$ matrix in this case. This data is generated at the design time and needs to be obtained at periodic intervals. Let Φ be a vector obtained at a given time, then we define a function $e_k^2 = (\Theta_k - \Phi) (\Theta_k - \Phi)^t$. Using this we generate another function $J_k = \log(e_k^2)$.

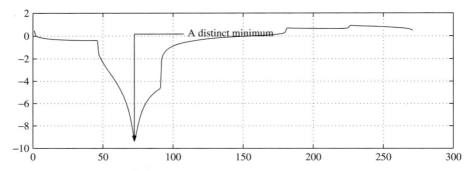

Figure 6.2 Minima in the direction of the emitter $J(k) = \log(e_k^2)$

This is plotted and shown in Figure 6.2, which has a distinct minimum representing the direction of the radiating source. Numerically locating this minimum gives the direction of the emitter.

6.3 Electronic Rotating Elements

In the previous section we saw that the phase delay φ between two elements can be modulated by rotating one element around the other. The rotation need not be done mechanically. To see how the rotating measuring system works, consider 12 elements positioned around a centre element (Figure 6.3). The delay at the ith element is

$$\varphi_i = \frac{2\pi}{\lambda} \hat{d}_i \cos(\Omega t - \theta) \quad \text{where} \quad \hat{d}_i = \frac{D}{2} \cos(\Delta i), \tag{6.1}$$

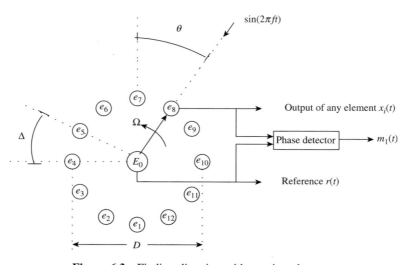

Figure 6.3 Finding direction with rotating elements

where $\hat{d}_i$ is the projected element distance and Ω is the angular rotation of the elements.

$$x_i(t) = \sin[2\pi f t + \varphi_i]. \tag{6.2}$$

The phase detector output $m_1(t)$ is obtained from $r(t) = \sin(2\pi f t)$ and $x_i(t)$:

$$m_1(t) = \cos(\varphi_i) = \cos\left[\left(\frac{\pi}{\lambda}\right)(D\cos\Delta i)\cos(\Omega t - \theta)\right]. \tag{6.3}$$

Signal $m_1(t)$ is passed through an FM discriminator and another phase detector, giving

$$\tilde{x}(t) = A_x \cos(\Omega t + \psi_1(t)), \tag{6.4}$$
$$\tilde{y}(t) = A_y \cos(\Omega t + \psi_2(t)), \tag{6.5}$$

where $A_x = K_x \cos(\theta)$, $A_y = K_y \sin(\theta)$ and $e = K_x - K_y$ is zero in a least-squares sense for a given system. Traditional analogue systems present DoA by inputting $\tilde{x}(t)$ and $\tilde{y}(t)$ to an X–Y scope, shown in Figure 6.4(a). The phase distortions $\psi_1(t)$ and $\psi_2(t)$ are the undesired phase modulations due to multiple reflections, instrumentation errors, etc., and the plot between $\tilde{x}(t)$ and $\tilde{y}(t)$ results in a distorted *figure of eight*. The ratio between the amplitudes of $\tilde{x}(t)$ and $\tilde{y}(t)$ has the DoA information in it and the sinusoids are in phase with each other. However, finding a direct ratio is catastrophic and here we formulate it as a DSP problem.

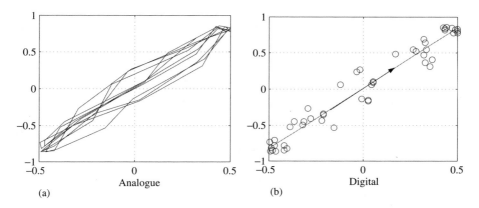

Figure 6.4   DOA using DSP in a rotary system

6.3.1 Problem Formulation

The two signals $\tilde{x}(t)$ and $\tilde{y}(t)$ are converted to the digital domain as

$$x_k = A_x \cos(\Omega k \, \delta t + \psi_k^1),$$
$$y_k = A_y \cos(\Omega k \, \delta t + \psi_k^2), \tag{6.6}$$

where δt is the sampling time.

6.3.2 Finding DoA

The prime objective here is to estimate the DoA (θ) and the quality of the DoA from the digitised signals x_k and y_k of N samples. The principle of the method is to plot x_k, y_k values in a 2D space and to find a best-fit straight line using *linear regression*. The slope of this line gives tan(DoA), from which we compute DoA. The mean square error is calculated and is used as a measure of the quality of the DoA given by the system. This is shown in Figure 6.4(b) and the theory is elaborated below.

6.3.3 Straight-Line Fit

The general equation for a straight line can be written as

$$y_k = mx_k + c \quad \text{for} \quad k = 1 \text{ to } N \tag{6.7}$$
$$= p_1 x_k + p_2, \tag{6.8}$$

where $m = p_1$ and $c = p_2$ are the slope and the abscissa of the straight line, respectively. Let

$$\mathbf{x}^t = [x_1, x_2, \ldots, x_N], \tag{6.9}$$
$$\mathbf{y}^t = [y_1, y_2, \ldots, y_N], \tag{6.10}$$
$$\mathbf{u}^t = [1, 1, \ldots, 1], \tag{6.11}$$
$$\underline{\mathbf{x}}_i^t = [x_i, 1]. \tag{6.12}$$

We formulate an $N \times 2$ rectangular matrix $\mathbf{A}$ given as

$$\mathbf{A} = [\mathbf{x}|\mathbf{u}], \tag{6.13}$$
$$\mathbf{p}^t = [p_1, p_2]. \tag{6.14}$$

Then we write the set of equations (6.8) in matrix form:

$$\mathbf{y} = \mathbf{A}.\mathbf{p} \tag{6.15}$$

Finding a best-fit straight line involves estimating $\mathbf{p}$ from the given $\mathbf{A}$ and $\mathbf{y}$. Premultiplying (6.15) by $\mathbf{A}^t$, we get

$$\mathbf{A}^t\mathbf{y} = (\mathbf{A}^t\mathbf{A})\mathbf{p}. \tag{6.16}$$

Here $(\mathbf{A}^t\mathbf{A})$ is a 2×2 matrix and if its inverse exists, we can write

$$\hat{\mathbf{p}} = \mathbf{A}^+\mathbf{y} \tag{6.17}$$

$$= (\mathbf{A}^t\mathbf{A})^{-1}\mathbf{A}^t\mathbf{y}, \tag{6.18}$$

where $\hat{\mathbf{p}}$ is the estimate of $\mathbf{p}$ corresponding to the best-fit straight line and $\mathbf{A}^+$ is known as the Moore–Penrose pseudo-inverse [1]. Equation (6.18) can be rewritten as

$$\hat{\mathbf{p}} = \left(\sum_{i=1}^{N}\mathbf{x}_i\mathbf{x}_i^t\right)^{-1}\left(\sum_{i=1}^{N}\mathbf{x}_i y_i\right) \tag{6.19}$$

$$= \begin{pmatrix} S_{xx} & S_x \\ S_x & N \end{pmatrix}^{-1}\begin{pmatrix} S_{xy} \\ S_y \end{pmatrix}, \tag{6.20}$$

where $\mathbf{x}_i$ is defined by (6.12) and the terms S_{xx}, S_x, S_y, S_{xy} are defined in (6.22) and (6.23).

But the above solution is computationally intensive as it involves matrix multiplications and a matrix inversion. A recursive solution is available for (6.19) but even though it is not as computationally intensive, it still requires considerable time [2, 3].

6.3.3.1 Closed-Form Solution

To avoid such time-consuming calculations, a closed-form solution for (6.20) is used [4]. The expressions are

$$\hat{p}_1 = \frac{(NS_{xy} - S_xS_y)}{(NS_{xx} - S_x^2)} \quad \text{and} \quad \hat{p}_2 = \frac{(S_y - \hat{p}_1S_x)}{N}, \tag{6.21}$$

where

$$S_x = \sum_{i=1}^{N}x_i \quad \text{and} \quad S_y = \sum_{i=1}^{N}y_i, \tag{6.22}$$

$$S_{xx} = \sum_{i=1}^{N}x_i^2 \quad \text{and} \quad S_{xy} = \sum_{i=1}^{N}x_iy_i. \tag{6.23}$$

Then the sum of squares of errors is given as SSE $= \| \mathbf{y} - \hat{\mathbf{y}} \|$, where $\hat{\mathbf{y}} = \mathbf{A}\hat{\mathbf{p}}$. So

$$
\begin{aligned}
\text{SSE} &= \| \mathbf{y} - \mathbf{A}\hat{\mathbf{p}} \| \\
&= \| \mathbf{y} - \mathbf{A}(\mathbf{A}^t\mathbf{A})^{-1}\mathbf{A}^t\mathbf{y} \| .
\end{aligned}
\tag{6.24}
$$

Again, to find the sum of squares of errors, a closed-form solution for (6.24) is used [4] to reduce the computational complexity:

$$
\text{SSE} = S_{yy} - \hat{p}_2 S_y - \hat{p}_1 S_{xy} \quad \text{where} \quad S_{yy} = \sum_{i=1}^{N} y_i^2.
\tag{6.25}
$$

We have illustrated various DSP techniques to improve or extract the hidden data by ingeniously combining ideas from systems engineering. The numbers of interest are $\hat{p}_1$, $\hat{p}_2$ (Figure 6.4) and SSE (6.25).

6.4 Summary

In this chapter we gave a detailed treatment of a specific problem in direction finding, showing the need to understand systems engineering in order to implement DSP algorithms. It is very difficult to isolate DSP problems from other areas of engineering. Practical problems tend to cover many areas but centre on one area that demands the principal focus.

References

1. A. Albert, *Regression and the Moore–Penrose Pseudoinverse*. New York: Academic Press, 1972.
2. D. M. Himmelblau, *Applied Non-linear Programming*, p. 75. New York: McGraw-Hill, 1972.
3. D. C. Montgomery *et al.*, *Introduction to Linear Regression*, Ch. 8. New York: John Wiley & Sons, Inc., 1982.
4. J. L. Devore, *Probability and Statistics for Engineers and the Sciences*. Monterey CA: Brooks Cole, 1987.
5. D. Curtis Scheler, *Introduction to Electronic Warfare*, pp. 326–30. Norwood MA: Artech House, 1990.
6. H. H. Jenkins, *Small-Aperture Radio Direction Finding*, pp. 139–48. Norwood MA: Artech House, 1991.
7. G. Multedo, 'Direction finding.' *Revue Technique Thomson-CSF*, **19**(2), 1987.

APPENDIX

MATLAB and C Programs

We consider this to be the most important part of the book. Each program was tested and has a comment statement to say which figures it generates; this is to put the reader at ease while running the program. Try to play with the programs by manipulating various parameters to see their effects. We have repeated the relevant chapter summary at the start of Section A.1 to A.6.

A.1 Chapter 1 MATLAB Programs

In this chapter we discussed the types of signals and the transformations a signal goes through before it is in a form suitable for input to a DSP processor. We introduced some common transforms, time and frequency domain interchanges, and the concepts of windowing, sampling and quantisation plus a model of an A/D converter. The last few sections give a flavour of DSP.

A.1.1 Program f1_1234

```
%Digital Signal Processing:A Practitioner's Approach
%Dr.Kaluri Venkata Ranga Rao,kaluri@ieee.org
%Generates figures 1.1,1.2
% Generations of Signals for 1st Chapter
% 25th July 2003
clear;clf;
% Narrow Band Signal
fc=0.3; wn_upper=fc+fc*0.01; wn_lower=fc-fc*0.01;
wn=[wn_lower wn_upper];
[b,a]=butter(2,wn);
t=1:1000; u=randn(size(t));
```

Digital Signal Processing: A Practitioner's Approach   K. V. Rangarao and R. K. Mallik
© 2005 John Wiley & Sons, Ltd

```
y=filter(b,a,u);
[sx,f]=spk(y,1);
subplot(221);plot(t,y);grid;
title('A')
xlabel(' Time ')
subplot(222);plot(f,sx);grid;
title('B')
xlabel(' Frequency ')
%title ('Fig 1.1');
pause;
%print -depsc f1_1;
%title('Fig 1.2');
%print -depsc f1_2;
pause;
% Broad band signal
fc=0.6; wn_upper=fc+fc*0.1; wn_lower=fc-fc*0.1;
wn=[wn_lower wn_upper];
[b,a]=butter(2,wn);
t=1:1000; u=randn(size(t));
y=filter(b,a,u);
[sx,f]=spk(y,1);
subplot(221);plot(t,y);grid;
title('A');
xlabel(' Time ')
subplot(222);plot(f,sx);grid;
title('B');
xlabel(' Frequency ')
pause;
%print -depsc f1_3;
%title('Fig 1.3')
%title ('Fig 1.4')
%print -depsc f1_4
```

A.1.2 Program f1_56

```
%Digital Signal Processing:A Practitioner's Approach
%Dr.Kaluri Venkata Ranga Rao, kaluri@ieee.org
%Generates figure 1.3

% Generations of Signals for 1st Chapter
% 25th July 2003
clear;clf;
% Narrow Multi Band Signals
fc=0.3; wn_upper=fc+fc*0.005; wn_lower=fc-fc*0.005;
```

```
wn=[wn_lower wn_upper];
[b,a]=butter(2,wn);
t=1:1000; u=randn(size(t));
y1=filter(b,a,u);
fc=0.1; wn_upper=fc+fc*0.1; wn_lower=fc-fc*0.1;
wn=[wn_lower wn_upper];
[b,a]=butter(2,wn);
t=1:1000; u=randn(size(t));
y2=filter(b,a,u);
fc=0.475; y3=0.1*sin(2*pi*fc*t);
y=y1+ y2 + y3;
[sx,f]=spk(y,1);
subplot(221);plot(t,y);grid;
xlabel('Time');title('A');
subplot(222);plot(f,sx);grid;
xlabel('Frequency');title('B');
pause
%print -depsc f1_5
%title ('Fig 1.5')
%title ('Fig 1.6')
%print -depsc f1_6
```

A.1.3 Program f1_78

```
%Digital Signal Processing:A Practitioner's Approach
%Dr.Kaluri Venkata Ranga Rao, kaluri@ieee.org
%Generates figure 1.4

% Generations of Signals for 1st Chapter
% 25th July 2003
clear;clf;
% Statistical Signal
t=1:5000; u=randn(size(t));
[x,y]=pdf_y(u,50);
yhat=((exp(-(x.*x)*0.5))/(sqrt(2*pi)));
scale=sum(yhat); yhat=yhat/scale;
subplot(221);plot(t,u);grid;
xlabel('Time');title('A');
subplot(222);plot(x,y,'o',x,y,x,yhat,'.-');grid;
xlabel(' Probability Density Function ');title('B');
pause;
print -depsc f1_7;
%title ('Fig 1.7')
```

```
%title ('Fig 1.8')
%print -depsc f1_8;
```

A.1.4 Program f11_901

```
%Digital Signal Processing:A Practitioner's Approach
%Dr.Kaluri Venkata Ranga Rao, kaluri@ieee.org
%Generates figure 1.5 and 1.6

    clear;close;
fs=10e3;T=1/fs;
t = 1:1000;
fc = 0.3;
Q = 100;f1 = fc - (fc/Q);f2 = fc + (fc/Q);
uk = randn(size(t));
wn = [f1 f2];
[b,a] = butter(2,wn);
[h,w] = freqz(b,a,360);
m = abs(h);F = w / (2*pi);
xk = filter(b,a,uk);
t=t*T;
subplot(211);plot(t,xk);grid;
% title(' Fig 1.9');
  print -depsc f1_9; pause;
  [sxk,f] = spk(xk,1);
subplot(211);plot(f*fs,sxk);grid;
% title(' Fig 1.10');
  xlabel(' Frequency in Hz. ');
  print -depsc f11_0; pause
subplot(211);plot(F*fs,m);grid;
% title(' Fig 1.11');
  xlabel(' Frequency in Hz. ');
  print -depsc f11_1
```

A.1.5 Program f11_23458

```
%Digital Signal Processing:A Practitioner's Approach
%Dr.Kaluri Venkata Ranga Rao, kaluri@ieee.org
%Generates figure 1.7,1.8,1.9
%This program needs spk.m, expand.m

clear;close;
fs=10e3;T=1/fs;
f=0.001;
```

```
t=linspace(0,1000,1000); a=0.3
x=sin(2*pi*f*t)+a*sin(2*pi*f*t*3);
impulse=ones([1 40]);t=t*T;
timp_train=expand(impulse,26);
imp_train=timp_train(1:1000);
x_smple=x.*imp_train;
x2=round((round(1000*x_smple))/200);
x2=x2/5;
subplot(221);plot(t,x,t,imp_train);grid;
    xlabel(' Seconds ');
%title(' Sampling of Continious signal ');
    %title(' Fig 1.12 ');
%subplot(221);plot(t,x_smple);grid;
    % xlabel(' Seconds ');
    title('A');
    subplot(222)
plot(t,x_smple,t,x,'-.');grid
    xlabel(' Seconds ') ;title( 'B');
%   print -depsc f11_4; pause
%   print -depsc f11_2; pause;
%   print -depsc f11_3;pause;
    clf
subplot(221);plot(t,x2,t,x,'-.');grid;
title(' A ');
%title(' Fig 1.18 ');
    subplot(222);plot(t,x_smple-x2);grid
    title(' B ');
    xlabel(' Seconds ');
[sx_smple,f1]=spk(x_smple,1);
    [sx,f]=spk(x,1);
%   print -depsc f11_8
    pause;
    clf
%subplot(221)
%plot(t,x) ;grid
%title (' Continious Signal ')
%title (' Fig 1.15 ')
%subplot (223)
%plot (f (1:10) *f s, sx(1:10));grid
%title(' Fig 1.16 ')
%xlabel(' frequency Hz ')
    subplot(211)
plot(f1*fs,sx_smple*25);grid
%title(' Fig 1.15 ')
```

```
    xlabel(' frequency Hz ')
    pause
%   print -depsc f11_5
%zoom;
```

A.2 Chapter 2 MATLAB Programs

This chapter was a refresher on some relevant engineering topics. It should help you with the rest of the book. We considered the autocorrelation function, representation of linear systems, and noise and its propagation in linear systems. We also discussed the need to know about systems reliability. Most problems in real life are inverse problems, so we introduced the Moore–Penrose pseudo-inverse. We concluded by providing an insight into binary number systems.

A.2.1 Program f2_345

```
%Digital Signal Processing:A Practitioner's Approach
%Dr.Kaluri Venkata Ranga Rao, kaluri@ieee.org
%Generates figures 2.3,2.4

clear;clf
delta=0.1;f=100;
omega=2*pi*f;
b=omega*omega;
a=[1 2*delta*omega omega*omega];
Ts=1/(10*f);
sysc=tf(b,a)
sysd=c2d(sysc,Ts,'tustin');
sysd1=c2d(sysc,2*Ts,'tustin');
[b1 a1]=tfdata(sysd,'v');
[b2 a2]=tfdata(sysd1,'v');
[y,tc]=impulse(sysc); tfinal=length(tc);
td=1:tfinal; x=td*0; x(1)=1;
y1=f ilter(b1,a1,x); y2=filter(b2,a2,x);
delt=1;n1=length(x);delf = (1/delt)/n1;
Sy = fft(y2);
M = abs(Sy)*delt; n=fix(length(M)/2);
[Mx i]=max(M(1:n));
mark=zeros(size(M)); mark(i)=M(i);
f1=1:n; f1=f1*delf;
Rtow=real(ifft(fftshift(M)));
Rtow=fftshift(Rtow);
T=1:length(Rtow); T=T-fix(length(Rtow)/2);
T=T*delt;
```

```
subplot(222);plot(f1,M(1:n),f1,mark(1:n));grid;title('B')
ylabel(' Power Spectrum ');
%title ( ' Fourier transform of R(tow) ')
%title ( ' Fig 2.5 ')
%print -depsc f2_5;pause;
subplot(221);plot(T,Rtow);grid;title('A')
ylabel(' Auto Correlation ');
%title ( ' Auto Correlation function R(tow) ');
%title ( ' Fig 2.4 ');
print -depsc f2_4;pause
td=td-1;
%subplot(211) ;plot(tc,y) ;grid
%subplot(212) ;plot(td,y1,td,y2);grid;pause;
%subplot(211);plot(td(1:60),y2(1:60),td(1:60),y2(1:60),'o');grid
subplot(211);stem(td(1:30), y2(1:30));grid
ylabel(' Impulse Response ');xlabel(' Time ')
%title('Fig 2.3');
print -depsc f2_3
```

A.2.2 Program f2_7

```
%Digital Signal Processing:A Practitioner's Approach
%Dr.Kaluri Venkata Ranga Rao, kaluri@ieee.org
%Generates figure 2.6

clear;close;
delta=0.1;
f=100;
omega=2*pi*f;
b1=omega*omega;
a1=[1 2*delta*omega omega*omega];
sysc=tf(b1,a1);
Ts1=1/(4*f);
Ts2=1/(50*f);
imax=10;
Ts3=linspace(Ts1,Ts2,imax);
for i=1:imax
Ts=Ts3(i);
sysd=c2d(sysc,Ts,'tustin');
[b a]=tfdata(sysd,'v');
q=abs(roots(a));
g1(i)=b(1);
r1(i)=q(1);
```

```
q=angle(roots(a));
theta1(i)=q(1);
end
imax=20;jmax=8;
delt=linspace(0.05,0.9,imax);
Ts1=1/(0.5*f) ;Ts2=1/(100*f);
Ts3=linspace(Ts2,Ts1,jmax);
Ts0=linspace(Ts3(1),Ts3(3),jmax);
Ts0(jmax+1:2*jmax-3)=Ts3(4:jmax);
jmax=length(Ts0);
for j=1:jmax
Ts=Ts0(j);
k=1;
for i=1:imax
delta=delt(i);
f=100;
omega=2*pi*f;
b1=omega*omega;
a1=[1 2*delta*omega omega*omega];
sysc=tf(b1,a1);
sysd=c2d(sysc,Ts,'tustin');
[b a]=tfdata(sysd,'v');
q1=abs(roots(a));
g2(i)=b(1);
r2(k)=q1(1);
q2=angle(roots(a));
theta2(k)=q2(1);
k=k+1;r2(k)=q1(2);theta2(k)=q2(2);k=k+1;
end
%p=polyfit(delt,r2,1);
%r2hat=p(1)*delt+p(2);
%m(j)=p(1);c(j)=p(2);
Tss(j)=Ts;
A(j,:)=delt;B(j,:)=r2;C(j,:)=theta2;
%subplot(211) ;plot(delt,r2, 'o', delt,r2hat) ;grid
%subplot (212) ;plot (delt, theta2* (180/pi), ' o') ;grid
%polar (theta2,r2, ' o');grid
%theta(j)=mean(theta2*(180/pi));
%Ts
end
subplot(111);polar(C,B,'o');grid;
%[xp,yp]=ginput(1);text(xp,yp,' Fig 2.7 ');
%[xp,yp]=ginput(1) ;text(xp,yp,'<--- Constant Sampling Rate ');
%[xp,yp]=ginput(1) ;text(xp,yp,' Sensivity of System ');
```

```
pause;
NN=(Tss/Ts1);
%NN=1./NN;
markx=linspace(0,180,180);marky=markx*0; marky(30)=1;
  marky(120)=1;
subplot(221)
plot(NN,B,markx/180,marky,'b');grid;xlabel('Sampling Frequency');
ylabel('Radial pole position');title('A');
subplot(222)
plot(NN,C*180/pi);grid;xlabel('Sampling Frequency');
ylabel('Pole angle') ;title('B');
%print -depsc f2_7
```

A.2.3 Program f2_7b

```
%Digital Signal Processing:A Practitioner's Approach
%Dr.Kaluri Venkata Ranga Rao, kaluri@ieee.org
%Generates figure 2.10

clear;close;
n_first=100;n_middle=200;
i=1:n_first;
x=exp(-0.05*i); y=x;
for i=1:n_middle
x(n_first+i)=x(n_first);
end
for i=1:n_first
x(n_first+n_middle+i)=y(n_first+1-i);
end
k=1:length(x);
z=x*0;z=z+0.001;
x=x+0.01; x=x*0.6;
p1=n_first; p2=(p1+n_middle); p3=length(x);
subplot(211);
plot(k(1:p1),x(1:p1),'o',k(p1+1:p2),x(p1+1:p2),
   '-.',k(p2+1:p3),x(p2+1:p3),'*',k,z)
pause
%print -depsc f2_7b
```

A.2.4 Program f2_8a

```
%Digital Signal Processing:A Practitioner's Approach
%Dr.Kaluri Venkata Ranga Rao, kaluri@ieee.org
%Generates figure 2.9
```

```
clear;
kmax=5000;bins=100;order=5;
k=1:kmax; g=9.81;
mu_v=50; sig_v=0.2;
mu_th=pi/3; sig_th=pi/30;
temp=randn(size(k));
v=temp*sig_v; v=v+mu_v;
temp=randn(size(k));
th=temp*sig_th; th=th+mu_th;
d=v.*v.*sin(2*th)/g;
[N,X]=hist(d,bins); pdf=N/kmax;
[P,S]=polyfit(X,pdf,order);
pdf_hat=polyval(P,X);
subplot(221)
plot(k,v);grid;title(' A ');ylabel(' V in m/sec ');
subplot(222)
plot(k,th*180/pi);grid;title(' B ');ylabel(' Theta Deg ');
subplot(212)
plot(X,pdf,'.',X,pdf_hat);grid;title('C');ylabel('Probability');
```

A.2.5 Program f2_89

```
%Digital Signal Processing:A Practitioner's Approach
%Dr.Kaluri Venkata Ranga Rao, kaluri@ieee.org
%Generates figure 2.14
%This program needs pdf_y.m

clear;close;
delta=0.1;f=100;
omega=2*pi*f;
b=omega*omega;
a=[1 2*delta*omega omega*omega];
Ts=1/(10*f);
sysc=tf(b,a)
sysd=c2d(sysc,Ts,'tustin');
[b1 a1]=tfdata(sysd,'v');
plot(impulse(sysd));grid;
a=0.1;b=0.8;
for i=1:6000
u(i)=weibrnd(a,b);
end
z=filter(b1,a1,u);
[x,y]=pdf_y(u,30);
[x1,y1]=pdf_y(z,20);
```

```
pause;
subplot(211);plot(u);grid;
%title(' Fig 2.8 ');
%print -depsc f2_8;pause;
subplot(211);plot(x,y,'*',x,y,'-',x1,y1,'.-',x1,y1,'o');grid;
%title(' Fig 2.9 ');
[xp,yp]=ginput(1);text(xp,yp,'<---    Weibull distribution ');
[xp,yp]=ginput(1);text(xp,yp,'<---    Normal distribution ');
%print -depsc f2_9
```

A.2.6 Program f21_0123

```
%Digital Signal Processing:A Practitioner's Approach
%Dr.Kaluri Venkata Ranga Rao, kaluri@ieee.org
%Generates figure 2.12,2.13
%This program needs pdf_y.m
% Generations of Signals for 2nd Chapter
% 25th July 2003
clear;clf;
% Statistical Signal
t=1:10000; u=randn(size(t));
[x,y]=pdf_y(u,50);
sigma=1;
xx=(x.*x)/(2*sigma);
yhat=exp(-xx)/(sqrt(2*pi*sigma));
scale=sum(yhat); yhat=yhat/scale;
[su,f1]=spk(u,1);
subplot(222);plot(x,y,'o',x,y,x,yhat,'.-');grid;
ylabel(' Statistics of uk ');
title(' B ')'
%title('Fig 2.10')
%print -depsc f21_0;pause;
subplot(221);
plot(f1,su);grid;
%title('Fig 2.11')
ylabel(' White Noise ');
title(' A ')'
%print -depsc f21_0;pause;
%print -depsc f21_1
pause
delta=0.1;
%delta=0.01;
f=100;
```

```
omega=2*pi*f;
b=omega*omega;
a=[1 2*delta*omega omega*omega];
Ts=1/(10*f);
sysc=tf(b,a);
sysd=c2d(sysc,Ts,'tustin');
[b1 a1]=tfdata(sysd,'v');
y1=filter(b1,a1,u);
%pause;
[x,y]=pdf_y(y1,50);
[sy1,f1]=spk(y1,1);
gain=mean(sy1(1:150));
N=length(f1); m=abs(freqz(b1,a1,N));
sigma=1.5;
%sigma=12;
xx=(x.*x)/(2*sigma);
yhat=exp(-xx)/(sqrt(2*pi*sigma));
scale=sum(yhat); yhat=yhat/scale;
subplot(222);plot(x,y,'o',x,y,x,yhat,'.-');grid;
ylabel(' PDF of yk ');title(' B ');
%title('Fig 2.12')
subplot(221);plot(f1,sy1,'.',f1,m*gain,'.-');grid;
ylabel(' Coloured Noise yk ');title(' A ');
%title('Fig 2.13')
%xlabel(' Coloured Gaussian Noise - Ouput')
%print -depsc f21_2;pause;
%print -depsc f21_3
```

A.2.7 Program f21_67

```
%Digital Signal Processing:A Practitioner's Approach
%Dr.Kaluri Venkata Ranga Rao, kaluri@ieee.org
%Generates figures 2.17,2.18

clear;close
sigma_x=0.2;
sigma_y=0.2;
x_val=-5;y_val=7;
row_max=8;
A=fix((rand([row_max 2])-0.5)*20);
x1=x_val-randn( [1 row_max])*sigma_x;
x2=y_val-randn([1 row_max])*sigma_y;
c1=A(:,1); p1=c1.*x1';
c2=A(:,2); p2=c2.*x2';
```

```
p=fix(p1+p2);
x3=inv(A'*A)*A'*p;
ek=A*x3-p;
sigma=std(ek);
x=x_val-3:0.1:x_val+3;
for i=1:row_max
q=A(i,:);
y=-q(1)/q(2)*x+p(i)/q(2);
B(i,:)=x;C(i,:)=y;
end
N=length(x);
i=i+1;
x_h=linspace(min(min(B)),max(max(B)), N);
x_v=ones(size(x_h))*y_val;
y_v=linspace(min(min(C)),max(max(C)), N);
y_h=ones(size(y_v))*x_val;
theta=linspace(0,2*pi,N);
x=x3(1)-sigma*cos(theta);
y=x3(2)-sigma*sin(theta);
subplot(211);
plot(B',C','.',y_h,y_v,x_h,x_v,x,y,'o',x,y);grid;
% print -depsc f21_6
subplot(212)
stem(ek);grid
% print -depsc f21_7
```

A.3 Chapter 3 MATLAB Programs

This chapter described ways to specify a given filter. We presented a variety of popular filters and their characteristics. We looked at IIR filters with real coefficients and truncated IIR filters for use in FIR designs. We also described a bank of filters with complex coefficients pointing to discrete Fourier transforms. We gave a simple presentation of adaptive filters and inverse problems having practical significance. We demonstrated an application of BFSK demodulation and adaptive phase shifting. And we looked at a practical inverse problem from target tracking. We added an important dimension by giving Kalman filter applications. Then we concluded this big topic with an application from array processing.

A.3.1 Program f3_1

```
%Digital Signal Processing:A Practitioner's Approach
%Dr.Kaluri Venkata Ranga Rao , kaluri@ieee.org
%Generates figure 3.1
```

```
clear;close;
n=6;rdb=0.4;wn=0.4;
[b,a]=cheby1(n,rdb,wn);
%[b,a]=butter(n,wn);
N=360;
fn=linspace(0,0.5,N);
sysd=tf(b,a,1); hk=impulse(sysd);
m=abs(freqz(b,a,N));
phi=angle(freqz(b,a,N))*180/pi;
nmax=length(fn); n1=round(wn*nmax*1.1);
line_1=ones(size(fn(1:n1)))*0.92;
line_2=ones(size(fn(1:n1)))*1.02; part_1=fn(1: n1);
subplot(211);plot(fn,m,part_1,line_1,part_1,line_2);grid;
print -depsc f3_1
%temp1=abs(roots(a)) ;temp=angle(roots(a));
% [theta,i]=sort(temp);r=temp1(i);pause;
%subplot(111) ;polar(theta,r, 'o') ;grid;
%subplot(212) ;plot(fn,phi,fn,phi_new) ;grid;
%[xp, yp]=ginput(1);
text(xp,yp,'O<--- Non-linear Phase of IIR Filter ')
%[xp,yp]=ginput(1);
text(xp,yp,'O<--- Linear Phase of FIR filter ')
%print -depsc f3_8
```

A.3.2 Program f3_1a

```
%Digital Signal Processing:A Practitioner's Approach
%Dr.Kaluri Venkata Ranga Rao, kaluri@ieee.org
%Generates figure 3.2

clear;clg
k=1:200;f=0.02;x=sin(2*pi*f*k)+0.2*rand(size(k));
subplot(211);stem(k,x);grid
```

A.3.3 Program f3_2

```
%Digital Signal Processing:A Practitioner's Approach
%Dr.Kaluri Venkata Ranga Rao, kaluri@ieee.org
%Generates figure 3.9

clear;close;
xk=1;xk1=0;
for i=1:20
xk2=xk1;xk1=xk;xk=1.8*xk1-xk2;
```

```
y(i)=xk;x(i)=i;
end
subplot(212); stem(x,y);grid
text(xp,yp,'1.8000   2.2400   2.2320   1.7776  0.9677-0.0358  -1.0321');
text(xp,yp,'-1.8220  -2.2475  -2.2235  -1.7548  -0.9351  0.0715 1.0639');
%[xp,yp] =ginput (1);
text(xp,yp,'1.8435   2.2544   2.2144   1.7315   0.9023   -0.1073');
%[xp,yp]=ginput(1) ;text(xp,yp,' Fig 3.1 ')
print -depsc f3_2
```

A.3.4 Program f3_3

```
%Digital Signal Processing:A Practitioner's Approach
%Dr.Kaluri Venkata Ranga Rao, kaluri@ieee.org
%Generates figure 3.10

clear;close;
r=0.95;f=0.2;
a=[1 -2*r*cos(2*pi*f) r*r] ;
b=[1 0 -1];b=b*(1-r*r)*0.5;N=360;
fn=linspace(0,0.5,N);
sysd=tf(b,a,1);
m=abs(freqz(b,a,N));
phi=angle(freqz(b,a,N))*180/pi;phi=phi/180;
subplot(211);plot(fn,m,fn,phi);grid;
[xp,yp]=ginput(1);text(xp,yp,'<--- Phase +90 to -90 deg ')
[xp,yp]=ginput(1);text(xp,yp,'<--- Amplitude Spectrrum ')
print -depsc f3_3
```

A.3.5 Program f3_45

```
%Digital Signal Processing:A Practitioner's Approach
%Dr.Kaluri Venkata Ranga Rao, kaluri@ieee.org
%Generates figures 3.11,3.12

clear;close;
N=50;
fn=linspace(0,0.5,N);
rmin=0.3;rmax=0.95;
R=linspace(rmin,rmax,N);
for k=1:N
r=R(k);
f=0.2;
p=2*r*cos(2*pi*f);
```

```
a1=[1 -r*p r*r];
b1=[r*r -r*p 1];
phi=angle(freqz(b1,a1,N))*180/pi;
i=find(phi > 0 );phi(i)=phi(i)- 360;
A(k,:)=phi';
B(k,:)=fn;
end
subplot(211);
plot(B,A,B(1,:),A(1,:),B(1,:),A(1,:),'o',
B(N,:),A(N,:),B(N,:),A(N,:),'d');grid;
[xp,yp]=ginput(1);text(xp,yp,'<--- r = 0.3');
[xp,yp]=ginput(1);text(xp,yp,'<--- r = 0.95');
print -depsc f3_5
pause;
rmin=0.3;rmax=0.95;
R=linspace(rmin,rmax,N);
for k=1:N
r=R(k);f=0.2;
p=2*r*cos(2*pi*f);
a2=[1 -r];
b2=[-r 1];
phi=angle(freqz(b2,a2,N))*180/pi;
A(k,:)=phi';B(k,:)=fn;
end
subplot(211)
plot (B,A,B(1,:),A(1,:),B(1,:),A(1,:),'o',
B(N,:),A(N,:),B(N,:),A(N,:),'d');grid;
[xp,yp]=ginput(1);text(xp,yp,'<--- r = 0.3');
[xp,yp]=ginput(1);text(xp,yp,'<--- r = 0.95');
print -depsc f3_4
```

A.3.6 Program f3_6

```
%Digital Signal Processing:A Practitioner's Approach
%Dr.Kaluri Venkata Ranga Rao, kaluri@ieee.org
%Generates figure 3.17

clear;close;
r=0.95;f=0.2;p=2*cos(2*pi*f);g=(1-r*r)*0.5;
a=[1 -r*p r*r];
b=[(1+r*r)/2 -r*p (1+r*r)/2];
%b=b*(1-r*r)*0.5;
N=360;
fn=linspace(0,0.5,N);
```

```
m=abs(freqz(b,a,N));
phi=angle(freqz(b,a,N))*180/pi;
i=find( phi > 0 ); phi(i)=phi(i) - 360;
phi1=((phi/180) + 1)*0.25+0.5;
subplot(211);plot(fn,m,fn,phi1);grid;
[xp,yp]=ginput(1);text(xp,yp,'<--- Phase 0 to -360 deg ')
[xp,yp]=ginput(1);text(xp,yp,'<--- Amplitude Spectrrum ')
print -depsc f3_6
```

A.3.7 Program f31_4

```
%Digital Signal Processing:A Practitioner's Approach
%Dr.Kaluri Venkata Ranga Rao, kaluri@ieee.org

clear;close;
n=6;rdb=0.4;wn=0.4;
[b,a]=cheby1(n,rdb,wn);
%[b,a]=butter(n,wn);
N=360;
fn=linspace(0,0.5,N);
sysd=tf(b,a,1); hk=impulse(sysd);
m=abs(freqz(b,a,N));
phi=angle(freqz(b,a,N))*180/pi;
nmax=length(fn); n1=round(wn*nmax*1.1);
line_1=ones(size(fn(1:n1)))*0.92;
line_2=ones(size(fn(1:n1)))*1.02; part_1=fn(1: n1);
subplot(211);plot(fn,m,part_1,line_1,part_1,line_2);grid;
temp1=abs(roots(a));
temp=angle(roots(a)); [theta,i]=sort(temp);r=temp1(i);
subplot(111);polar(theta,r,'o');grid;
x=r.*sin(theta);y=r.*cos(theta);
[p,s]=polyfit(x,y,2);X=linspace(max(x), min(x),30);
Y=polyval(p,X); theta_x=atan(X./Y);r_x=sqrt(X.*X+Y.*Y);
hold;
polar(theta_x,r_x,'-.');
pause;
%xlabel('Chebyshev Filter');
%[xp,yp]=ginput (1) ;text (xp,yp,' Chebyshev Filter ')
print -depsc f31_4
%temp1=abs (roots (b));
temp=angle(roots(b));[theta,i]=sort(temp);r=temp1(i);
%pause;
%subplot(111) ;polar(theta,r, 'o') ;grid;
%subplot (212) ;plot (fn,phi,fn,phi_new) ;grid;
```

```
%[xp,yp]=ginput(1) ;text(xp,yp, 'O<--- Non-linear Phase of IIR Filter ')
%[xp,yp]=ginput(1) ;text(xp,yp,'O<--- Linear Phase of FIR filter ')
%print -depsc f3_8
```

A.3.8 Program f31_5

```
%Digital Signal Processing:A Practitioner's Approach
%Dr.Kaluri Venkata Ranga Rao, kaluri@ieee.org
%Generates figure 3.18,3.19

clear;close;
n=6;rdb=0.4;wn=0.4;
[b,a]=butter(n,wn);
N=360;
fn=linspace(0,0.5,N);
sysd=tf(b,a,1); hk=impulse(sysd);
m=abs(freqz(b,a,N));
phi=angle(freqz(b,a,N))*180/pi;
nmax=length(fn); n1=round(wn*nmax*1.1);
line_1=ones(size(fn(1:n1)))*0.92;
line_2=ones(size(fn(1:n1)))*1.02; part_1=fn(1: n1);
subplot(211);plot(fn,m,part_1,line_1,part_1,line_2);grid;
temp1=abs(roots(a));
temp=angle(roots(a)); [theta,i]=sort(temp);r=temp1(i);
subplot(111);polar(theta,r,'o');grid;
x=r.*sin(theta);y=r.*cos(theta);
[p,s]=polyfit(x,y,2);X=linspace(max(x), min(x),30);
Y=polyval(p,X); theta_x=atan(X./Y);r_x=sqrt(X.*X+Y.*Y);
hold;
polar(theta_x,r_x,'-.');
%xlabel('Butterworth Filter');
%[xp,yp]=ginput(1);text(xp,yp,'Butterworth Filter ')
%[xp,yp]=ginput(1);text(xp,yp,'<--- 2nd Order pole trace ')
pause;
%temp1=abs (roots (b));
temp=angle(roots(b));[theta,i]=sort(temp);r=temp1(i);
%pause;
%subplot(111) ;polar(theta,r, 'o') ;grid;
%subplot (212) ;plot (fn,phi,fn,phi_new) ;grid;
%[xp,yp]=ginput(1);
%text(xp,yp, 'O<--- Non-linear Phase of IIR Filter ')
%[xp, yp]=ginput(1);
%text(xp,yp, 'O<--- Linear Phase of FIR filter ')
```

A.3.9 Program f31_6a

```
%Digital Signal Processing:A Practitioner's Approach
%Dr.Kaluri Venkata Ranga Rao, kaluri@ieee.org
%Generates figure 3.16

clear;close;
t=1:500;
f=0.15; x=sin(2*pi*f*t);
phase=-30; phase=phase*pi/180;
z=sin(2*pi*f*t+phase);
plot(x,y);grid
R=linspace(0.7,0.97,200); n=length(R);
for i=1:n
r=R(i);
a=[1 -r];b=[1 -1/r];
y=filter(b,a,z);
Jk(i)=sum((x-y).*(x-y));
%plot(t(1:20),x(1:20),'o',t(1:20),y(1:20),'*');grid;
%pause;
end
subplot(211)
plot(R,log(Jk));
xlabel('r');
ylabel('Jk(r)');grid
%pause
%print -depsc f31_6a
```

A.3.10 Program f31_78

```
%Digital Signal Processing:A Practitioner's Approach
%Dr.Kaluri Venkata Ranga Rao, kaluri@ieee.org
%Generates figure 3.3,3.4

clear;close;
r=0.95;f=0.2;
a=[1 -2*r*cos(2*pi*f) r*r] ;
b=[1 0 -1];b=b*(1-r*r)*0.5;N=360;
fn=linspace(0,0.5,N);
sysd=tf(b,a,1); hk=impulse(sysd);nmax=8;
m=abs(freqz(b,a,N));
phi=angle(freqz(b,a,N))*180/pi;
bnew=hk(nmax:-1:1);
bnew(nmax+1: 2*nmax-1)=hk(2:nmax);
```

```
phi_new=angle(freqz(bnew,1,N))*180/pi;
subplot(221);
stem(bnew);grid;title('A');ylabel('hk');
subplot(222);plot(fn,phi,fn,phi_new);grid;title('B');ylabel('deg');
pause;
%print -depsc f3_7;
m_new=abs(freqz(bnew,1,N));m_new=m_new/(max(m_new));
subplot(211);plot(fn,m,fn,m_new);grid;
pause;
%print -depsc f3_8;pause;
```

A.3.11 Program f3_789

```
%Digital Signal Processing:A Practitioner's Approach
%Dr.Kaluri Venkata Ranga Rao, kaluri@ieee.org
%Generates figure 3.21,3.22

clear
close
[m1,f]=freq(0.5);
[m2,f]=freq(0.6);
[m3,f]=freq(0.7);
[m4,f]=freq(0.8);
[m5,f]=freq(0.9);
[m6,f]=freq(0.95);
[m7,f]=freq(0.98);
mn=[m1,m2,m3,m4,m5,m6,m7];
[m11,f]=freqt(36);
[m21,f]=freqt(54);
[m31,f]=freqt(72);
[m41,f]=freqt(90);
[m51,f]=freqt(108);
[m61,f]=freqt(126);
[m71,f]=freqt(144);
m12=[m11 m21 m31 m41 m51 m61 m71];
subplot(211)
plot(f,m12);
xlabel ('Normalised Frequency');
ylabel ('Magnitude')
grid;
%gtext('theta = 36');
%gtext('theta = 144');
  pause
  %print -depsc f31_8
```

```
subplot(211)
plot(f,mn);
xlabel('Normalised Frequency')
ylabel('Magnitude')
grid;
 pause;
 %print -depsc f31_7
%gtext('r = 0.98');
```

A.3.12 Program f31_90

```
%Digital Signal Processing:A Practitioner's Approach
%Dr.Kaluri Venkata Ranga Rao, kaluri@ieee.org
%Generates figures 3.23,3.24

clear
close
range=2;
N=200;
delp=range/N;
p=-1;
r=0.95;
r1=0.5*(1-r*r);
k=1:150;
f=0.2;
power=0.5;
u=sin(2*pi*f*k);
noise=power*randn(size(u));
u=u+0*noise;
for i=1:N
 b1=[r1 0 -r1];
 b2=[0 2*r];
 a=[1 -2*r*p r*r];
 p=p+delp;
 fhat=acos(p)/(2*pi);
 x=filter(b1,a,u);
 sk=filter(b2,a,x);
       ek = x - u;
 amp(i) = mean(ek.*ek);
%   F(i)=p;
 F(i)=fhat;
       slope(i)=2*mean(ek.*sk);
end
subplot(211)
```

```
plot(F,amp);
grid;
xlabel p
ylabel J(p)
%title (' Objective function for r=0.95 at 3dB SNR ');
pause
  %print -depsc f31_9
subplot(211)
plot(F,slope);
grid;
xlabel p
ylabel ( 'slope of [J(p)]')
  pause
  %print -depsc f32_0
%zoom
```

A.3.13 Program f31_9a

```
%Digital Signal Processing:A Practitioner's Approach
%Dr.Kaluri Venkata Ranga Rao, kaluri@ieee.org
%Generates figure 3.29

clear;clf;
b1=[1 0.6]; a1=[1 -1.7 1.53 -0.648];
[h,w]=freqz(b1,a1,180);m1=abs(h);f=w/(2*pi);
subplot(221);
plot(f,m1);grid;title('A');xlabel('Frequency');
b2=0.999317; a2=[1 -2.16988 2.41286 -1.47436 0.365412];
[h,w]=freqz(b2,a2,180);m2=abs(h);f=w/(2*pi);
subplot(222);
plot(f,m2,f,m1,'.-');grid;title('B');xlabel('Frequency')
%pause;
%print -depsc f31_9a
```

A.3.14 Program f3_6_1a

```
%Digital Signal Processing:A Practitioner's Approach
%Dr.Kaluri Venkata Ranga Rao, kaluri@ieee.org
%Generates figure 3.31

clear;close;
vw=15;N=2000;
yw_max=N;yw=0:yw_max;
delta_y=2*pi/yw_max;
xw=500*sin(delta_y*yw);
```

```
delta_t=0.25;
k=0:N;t=k*delta_t;
x_0=10000;y_0=8000;
v_T=23;theta=120*(pi/180);
x_T=x_0+v_T*cos(theta)*t;
y_T=y_0+v_T*sin(theta)*t;
noise=0.003*randn(size(t));
beta_m=atan2((x_T-xw),(y_T-yw))+noise;
sum1=ones([4 4]); sum1=sum1*0;  sum2=[0 0 0 0]';
for i=0:N
  u=[cos(beta_m(i+1)) -sin(beta_m(i+1)) t(i+1)*cos(beta_m(i+1))
    -t(i+1)*sin(beta)
  yk=xw(i+1)*cos(beta_m(i+1))-yw(i+1)*sin(beta_m(i+1));
  sum1=sum1 + u'*u;
  sum2=sum2 + u'*yk;
end
%%%%%%%%%%%%%%%%%%%%%%%%%%%%%%%%%%%%%%%%%%%%%%%%%%%%%%%%%
T=linspace(t(1), t(N+1),10);
X=inv(sum1)*sum2;
xhat_0=X(1);
yhat_0=X(2);
vhat=sqrt(X(3)*X(3)+X(4)*X(4));
xhat_T=xhat_0+vhat*cos(theta)*T;
yhat_T=yhat_0+vhat*sin(theta)*T;
subplot(221);
plot(xw/1000,yw/1000,x_T/1000,y_T/1000,xhat_T/1000,yhat_T/
  1000,'.-');grid;
title('A'); xlabel(' Target & Watcher '); ylabel(' Km ')
subplot(222);plot(t,beta_m*180/pi);grid
title('B');xlabel('seconds');ylabel(' beta-m in deg ');
pause;
```

A.3.15 Program f3_6_5

```
%Digital Signal Processing:A Practitioner's Approach
%Dr.Kaluri Venkata Ranga Rao, kaluri@ieee.org
%Generates figure 3.32

% generate a linear reggression graph
clear;close;
t=linspace(0,2*pi,100);
  x=cos(t); y=x+0.1*randn(size(t));
  subplot(211);
  plot(t,x,t,y,'o');grid;
```

```
xlabel(' Time ');
ylabel(' angle ');
pause
%  print -depsc f6_5
```

A.3.16 Program f3_6_9

```
%Digital Signal Processing:A Practitioner's Approach
%Dr.Kaluri Venkata Ranga Rao, kaluri@ieee.org
%Generates figure 3.34

clear;clf;
nmax=2000;
t=1:nmax+9; f=0.05;
xreal(1:1000)=2*sin(2*pi*f*t(1:1000));
xreal(1001:2000)=2*sin(2*pi*f*t(1001:2000));
theta=2*pi*f;
% plot(u);grid;pause;
% plot(xreal);pause;
noise=randn(size(xreal));
y=xreal+0.5*noise;
a=[2*cos(theta) -1; 1 0];
b=[0 0]'; u=0;
c=[1 0];
  x_prev=[.01. 01]';
p_prev=[10^6 0 ; 0 10^6];
N=0; s=100;
for i=1:s
N=N+1;
pred_x = a*x_prev+b*u;
  pred_p = a*p_prev*a';
  err = c*pred_x - y(i);
  Y(i)=y(i)
arr(i)=err;
  xhat = pred_x - ((pred_p*c')/(1+c*pred_p*c'))*err;
phat = pred_p-(pred_p*c'*c*pred_p)/(1+c*pred_p*c');
Xfilter(i)=xhat(1);
x_prev=xhat;
p_prev=phat;
  end
k=1:N;
% plot(k,xreal(1:N), k,y(1:N),k,Xfilter);grid;
  subplot(211)
plot(k,Y,'o',k,Xfilter);grid;
```

```
pause
print -depsc f6_9
```

A.3.17 Program f3_61_01

```
%Digital Signal Processing:A Practitioner's Approach
%Dr.Kaluri Venkata Ranga Rao, kaluri@ieee.org
%Generates figures 3.26,3.25
%This program needs rbn_sig.m and expand.m files.

clear;
close;
%N=5000;
% generate BFSK data
  xx=sign(randn([1 200]));
  fsk=rbn_sig(150,xx);
  ii=find(fsk==0);
  fsk=fsk*0.27;fsk(ii)=fsk(ii)+0.23;N=length(fsk);
  kk=1:N; z=sin(2*pi*fsk.*kk);
  [sz,ff]=spk(z,1);qq=800;
subplot(221);
plot(kk(1:qq), z(1: qq)*0.2+1.1,kk(1:qq),fsk(1:qq)*3);grid;
  title('A');xlabel(' Time subplot(222);plot(ff,sz);grid;
  title('B');xlabel(' Frequency ');
  pause;
% load fsk_data
% K=1:N;
% z=sin(2*pi*f*K);
noise=0.2236*randn(size(z));
z=z+0.1*noise;
lambda=0.9;
mu =0.002;
r=0.9;
vk=0;
delp =0;
N=5000;
range_p = 4;
delP = range_p/N;
xk1=0; xk=0;
sk1=0; sk=0;
uk1=0; uk=0;
P = 0;
for k=1:N
uk2=uk1; uk1=uk;
```

```
uk=z(k);
  xk2=xk1; xk1=xk;
xk=r*p*xk1-r*r*xk2+(uk-uk2)*(1-r*r)*0.5;
  sk2=sk1; sk1=sk;
  sk=r*p*sk1-r*r*sk2+r*xk;
ek=xk-uk;
vk = vk*(lambda) + (1-lambda)*(ek*ek);
grad = 2*ek*sk1;
m(k)=mu;
if (k > 1)
delp = mu*grad/(vk+0.0001);
end
p = p - delp;
ref(k)=acos(0.5*p)/(2*pi);
end
subplot(211)
plot(kk(1:N),fsk(1:N),kk(1:N),ref,'-.')
grid
xlabel 'Sample Number'
ylabel 'p'
%title ' BFSK Demodulated Signal '
  pause;
  %print -depsc f61_1
```

A.3.18 Program f3_61_4

```
%Digital Signal Processing:A Practitioner's Approach
%Dr.Kaluri Venkata Ranga Rao, kaluri@ieee.org
%Generates figure 3.35

clear;clf;
t=linspace(0,2*pi,360);
k=linspace(0,2*pi,60);
z=sin(t);y=sin(k)+0.1*randn(size(k));
plot(t,z,k,y,'o');
% Use this and Paint and PPT to generate f61_4
```

A.3.19 Program f32_6a

```
%Digital Signal Processing:A Practitioner's Approach
%Dr.Kaluri Venkata Ranga Rao, kaluri@ieee.org
%Generates figure 3.28

clear;clg;
```

```
N=77; f=1/N; nmax=2000; K=1:nmax; phi=pi/20;
delay=40;
dly=1:delay; dly=dly*0;
bufer=1: delay+1; bufer=bufer*0;
Uk=cos(2*pi*f*K + phi ); rk=0.2; uk0=0; xk0=0; pk=0;
noise=randn(size(Uk));
Uk=Uk+0.0*noise;
loopGain=0.008;shift=0;
for k=1:nmax;
uk1=uk0;
xk1=xk0;
uk0=Uk(k);
xk0=rk*xk1 - rk*uk0 + uk1;
Xk(k)=xk0;
zk=uk0*xk0;
bufer(2:delay+1)=dly; bufer(1)=zk;
dly=bufer(1:delay);
ek=(dly(delay)-dly(1))/(delay);
pk=pk+ek; Pk(k)=pk;
rk=rk - loopGain*(pk+shift);
Rk(k)=rk; T(k)=k;
if abs(rk) > 0.95
rk=0.95;
end
end
subplot(221)
plot(Xk,Uk, '. ');grid; xlabel(' xk '); ylabel(' uk ');title(' A ');
subplot(222)
plot(T,Rk);grid;ylabel(' rk '); title(' B ');
```

A.4 Chapter 4 MATLAB Programs

In this chapter we used a matrix approach to explain the principles of the FFT. We did not used butterfly diagrams. Circular convolution was explained in depth. We presented a hardware scheme for implementation in real time. We looked at a problem of estimating frequency using a DFT. We elaborated a hardware structure for real-time implementation of continuous spectrum update. We ended by covering the principles of the network analyser, an example from RF systems.

A.4.1 Program f4_4

```
%Digital Signal Processing:A Practitioner's Approach
%Dr.Kaluri Venkata Ranga Rao, kaluri@ieee.org
%Generates figure 4.10
```

```
%This program needs p2f1.m, pulse.m and forier.m

% demo package written by Dr K.V.Rangarao
% this computes theoritical
% Fourier Series Coefficients of a rectangular
% pulse and compares with the FFT generated
% Fourier Series Coefficients.
clear;clf
tow=0.1; T=1.2; f1=1/T; A=1.3;mm=9
N=2^mm; x=pulse(tow,T,N); x=x*A;delt=1e-3;
n1=round(N/2);
[f,a,b]=forier(x,delt);
[a1,b1,F]=p2f1(tow,f1,A,N);
 %subplot (211);plot(x);grid;
 [aa ii]=max(a); s=ii-20;e=ii+20;
subplot(221);plot(f(s:e),a(s:e),'o',f(s:e),a1(s:e),'.-');
 grid;title('A')
 ylabel('ak');xlabel('Co-sine');
subplot(222);plot(f(s:e),b(s:e),'o',f(s:e),b1(s:e),'.-');
 grid;title('B')
 ylabel('bk');xlabel('sine');
 %print -depsc f4_4
```

A.4.2 Program f4_56

```
%Digital Signal Processing:A Practitioner's Approach
%Dr.Kaluri Venkata Ranga Rao, kaluri@ieee.org
%Generates figure 4.11,4,12

clear;close;
nx=50;ny=15;
kx=1:nx; ky=1:ny; f=0.05;
x=sin(2*pi*f*kx);
y=cos(2*pi*f*ky);
z=conv(x,y);t=1:length(z);
%subplot (211);
%plot (t,z);grid
m=nx-ny;
pad=ones([1 m])*0;
y(ny+1:nx)=pad;
zhat=real(ifft(fft(x).*fft(y)));
t1=1:length(zhat);
subplot(221)
plot(t,z,'.-',t1,zhat,'o');grid;nn=length(zhat);
e=z; e(1: nn)=e(1:nn)-zhat;
```

```
subplot(222)
plot(t,e,'.-');grid
pause;
%print -depsc f4_5
N=nx+ny-1;
m=N-ny;
pad=ones([1 m])*0;
y(ny+1:N)=pad;
m=N-nx;
pad=ones([1 m])*0;
x(nx+1:N)=pad;
zhat=real(ifft(fft(x).*fft(y)));
t1=1:length(zhat);
subplot(211)
plot(t,z,'.-',t1,zhat,'o');grid;nn=length(zhat)
e=z; e(1: nn)=e(1:nn)-zhat;
%subplot(222)
%plot(t,e,'.-');grid
%pause
%print -depsc f4_6
```

A.4.3 Program f4_62_0

```
%Digital Signal Processing:A Practitioner's Approach
%Dr.Kaluri Venkata Ranga Rao, kaluri@ieee.org
%Generates figure 4.18

  clear
  clg
  subplot(111)
f=0.01;
a=[1 -0.89944 0.404496];
b=[0.126 0.252 0.126];
t=1:2^11;
zi=[0 0];
%rand('normal');
   Fmax=50;
   F=linspace(0.05,0.37,Fmax);
   Kmax=length(F);
for k=1:Kmax;
  f=F(k);
x=cos(2*pi*f*t);
[y,zf]=filter(b,a,x,zi);
zi=zf;
```

```
[y,zf]=filter(b,a,x,zi);
l=length(x);
xx=randn(size(t));
y=y+0.5*xx;
win=hamming(l);
win=win';
sx=fft(x.*win);
sy=fft(y.*win);
mx=(abs(sx))/(l);
my=(abs(sy))/(l);
  phsex=atan2(imag(sx),real(sx))*180/pi;
  phsey=atan2(imag(sy),real(sy))*180/pi;
%phsex=(angle (sx)) *180/pi;
%phsey=(angle (sy)) *180/pi;
[mx1,i1]=max(mx);
[my1,i2]=max(my);
gain=my1/mx1;
%phx(k)=phsex(i1);
%phy(k)=phsey(i2);
phse3=phsey(i2)-phsex(i1);
mag(k)=gain;
phase(k)=phse3;
f1(k)=f;
%h1=samplz(b,a,f);
  Z=exp(sqrt(-1)*2*pi*f);
  h1=polyval(b,Z)./polyval(a,Z);
mag_true(k)=abs(h1);
phse_true(k)=(angle(h1))*(180/pi);
%phse_true(k)=(atan2(imag(h1),real(h1))) * (180/pi);
%f=f+0.04
end
%phz=phx-phy;
%save fdls.dat f1 mag mag_true phase phse_true
%plot (f1,phy,f1,phx,f1,phz); grid
%pause
subplot(221);
plot(f1,mag,'o',f1,mag_true,'.-');grid;title(' A ');
subplot(222);
plot(f1,phase,'o',f1,phse_true,'.-');grid;title(' B ');
%  FDLS starts from here
    imax=length(f1);
  K=0:10;
  for i=1:imax-8;
  for k=3:-1:1
```

```
    uk=cos(2*pi*f1(i)*k);
%   yk=mag_true(i)*cos(2*pi*f1(i)*k+phse_true(i)*pi/180);
    yk=mag(i)*cos(2*pi*f1(i)*k+phase(i)*pi/180);
    temp(4-k)=yk;
    temp(7-k)=uk;
    Uk=cos(2*pi*f1(i)*K);
    Yk=mag(i)*cos(2*pi*f1(i)*K+phase(i)*pi/180);
    end
%   subplot (211); stem (Uk);
%   subplot(212);stem(Yk);
    temp(2:3)=-temp(2:3);
    Y(i)=temp(1); A(i,:)=temp(2:6);
%   pause;
    end
    z=Y';
    p(1:2)=a(2:3); p(3:5)=b;
    phat=inv(A'*A)*A'*z;
```

A.4.4 Program f4_6_2

```
%Digital Signal Processing:A Practitioner's Approach
%Dr.Kaluri Venkata Ranga Rao, kaluri@ieee.org
%Generates figures 4.15,4.16

clear;clf;
w=20; T=100; f=0.3;records=6;
x=ones([1 T]);
x(w+1:T)=0*x(w+1:T);
for i=1:records
s=(i-1)*T+1; e=s+T-1;
z(s:e)=x;
end
t=1:length(z); noise=1.0*randn(size(t));
y=sin(2*pi*f*t)+noise;
puls=z.*y;
subplot(211);plot(t,puls);grid;
%title('A');
s1=1;e1=s1+w-1;
for i=1:records
s=(i-1)*T+1; e=s+w-1;
xnew(s1:e1)=puls(s:e);
s1=e1+1;e1=s1+w-1;
end
[sxnew,fx]=spk(xnew,1);
```

```
%subplot(222);plot(xnew);title('B');grid;
pause;
%print -depsc f6_2
subplot(221);plot(fx,sxnew,'.-');grid;title('A');xlabel(' Freq ');
[smax,i]=max(sxnew);s=4;e=7;
y1=sxnew(i-s:i+e);f1=fx(i-s:i+e);
f2=linspace(f1(1),f1(length(f1)),100);
p=polyfit(f1,y1,3);y1hat=polyval(p,f2);
subplot(222);
stem(f1,y1);grid
hold;
%plot(f1,y1,'.-',f2,y1hat);grid;title('B')
plot(f2,y1hat,'.');grid;title('B');
grid;xlabel(' Freq ');
pause;
%print -depsc f6_2a
```

A.4.5 Program f4_61_5b

```
%Digital Signal Processing:A Practitioner's Approach
%Dr.Kaluri Venkata Ranga Rao, kaluri@ieee.org
%Generates figure 4.3

clear;close;clf;
range=12;
F_u=linspace(0.01,0.22,range);
for k=1:range
tiks=150; f_smp=200; f_u=F_u(k);
x=1:tiks; x=x*0;
x(tiks)=1;
for i=1:f_smp
s=1+(i-1)*tiks; e=s+tiks-1;
u(s:e)=x;
end
b=1;a=[1 -1];
z=filter(b,a,u); nz=length(z); z(nz)=z(nz-1);
z=(z/max(z))*0.5;
%z=(z/max(z))*0.25;
t=1:nz;
% Frequency Sweep
xk=cos(2*pi*z.*t);
% Unknown signal
zk=sin(2*pi*f_u*t)+1*randn( [1 nz]);
```

```
% Band pass filter
f_c=0.25; r=0.98;
p=2*cos(2*pi*f_c); g=(1-r*r)*0.5; a=[1 -r*p r*r] ; b=[1 0 -1]*g;
% . . . . . . . . . . . . . . . .
uk=xk.*zk;
yk=filter(b,a,uk);
for i=1:f_smp
s=1+(i-1)*tiks; e=s+tiks-1;
blk=yk(s:e); rsp(i)=mean(blk.*blk);
end
%F=linspace(0,0.5,f_smp);
F=linspace(0,1,f_smp);
mid=round(f_smp/2);
[rsp1,i1]=max(rsp(1:mid));
[rsp2,i2]=max(rsp(mid+1:f_smp));
%[rsp1,i1]=max(rsp);
pk=rsp*0; pk(i1)=rsp1;
pk(i2+mid)=rsp2;
d(k)=F(i2+mid)-F(i1);
%d(k)=F(i1);
subplot(221)
%plot(F(1:mid),rsp(1:mid),'.-',F(1:mid),pk(1:mid));grid;title('A');
plot(F,rsp,'.-',F,pk);grid;title('A');
pause
end
P=polyfit(d,F_u,1);
Fhat=P(1)*d+P(2);
subplot(222);
plot(d,F_u,'o',d,Fhat);grid;title(' B ');
%pause
%print -depsc f61_5b
```

A.5 Chapter 5 Programs

In this chapter we described various hardware elements that are frequently used to convert the given analogue signal into digital form. We implemented a second-order BPF in SIMULINK and in fixed-point hardware then compared the results. We looked at a hardware realisation scheme that led us to consider pipelining and its associated problems. We gave an in-depth discussion on various methods of pipelining FIR and IIR filters. Then we looked at a complex algorithm for estimating frequency based on time series data and implemented on the DSP5630X using C. To implement this algorithm using FPGAs, we chose a design using FIR filters.

A.5.1 MATLAB Program f5_7a

```
%Digital Signal Processing:A Practitioner's Approach
%Dr.Kaluri Venkata Ranga Rao, kaluri@ieee.org
%Generates figure 5.8

clear;
r2_o=0; r1_o=0;
R2_o=0; R1_o=0;
for i=1:1000
  if i == 1
    u=1;
  else
    u=0;
  end
R1_i=(R2_o+u);
R2_i=-0.9875*R1_o+1.608*R1_i;
R1_o=R1_i;
R2_o=R2_i;
r1_i=fix(fix((r2_o+2^16*u))/256);
r2_i=-252*r1_o+411*r1_i;
r1_o=r1_i;
r2_o=r2_i;
x(i)=r1_o; X(i)=R1_o;
end
t=1:length(x); mark=ones(size(t))*255;
m=abs(fft(x)); M=abs(fft(256*X)); e=length(M)
k=1:e/2;k=k/e;
scale=1/max(M);
m1=scale*m(1:e/2);M1=scale*M(1:e/2);
[a p]=max(M1);
subplot(211)
plot(k(p-50:p+50),m1(p-50:p+50),'-o',k(p-50:p+50),
  M1(p-50:p+50));grid
xlabel('Normalised Frequency');
ylabel('Response');
pause;
print -depsc f5_7a
```

A.5.2 MATLAB Program f51_1

```
%Digital Signal Processing:A Practitioner's Approach
%Dr.Kaluri Venkata Ranga Rao, kaluri@ieee.org
%Generates figure 5.11
```

```
clear;
i=1;
for phi=20:1:150
r=0.9; theta=phi*(pi/180); p=2*cos(theta);
a1=-r*p; a2=r*r;
P1=[1 a1 a2]; P2=[1 -a1 a1*a1-a2];
P3=conv(P1,P2);
a3=r^3*(2*p-p^3); a4=r^4*(p*p-1);
P4=[1 0 0 a3 a4];
P4-P3;
R=abs(roots(P2)); r1(i)=R(1);r2(i)=R(2);Theta(i)=phi;
%pause
i=i+1;
end
subplot(211)
plot(Theta,r1,'*',Theta,r2,'o');grid
%pause
%print -depsc f51_1
```

A.5.3 MATLAB Program f51_2

```
%Digital Signal Processing:A Practitioner's Approach
%Dr.Kaluri Venkata Ranga Rao, kaluri@ieee.org
%Generates figure 5.13

close;clear
kmax=5000; theta=20; f=theta/360;
t=1:kmax;
z=sin(2*pi*f*t)+0.2*randn(size(t));
a1=0.1; a2=0.1;
r=0.98;
v=0.0;
Hk=[1e-9 0 0
  0 1e-9 0
  0 0 1e-9];
delp_k_old=[0 0 0]';
delp_k= [0 0 0]';
xk=0; xk1=0.1; sk=0; sk1=0; sk2=0; xk_corkt=0;
zk=0; zk1=0;
del_a1=0; del_a2=0;
old_del_a1=0; old_del_a2=0;
alpha1=0.95; delx_k=0;
pk=[a1 a2 v]';
```

```
yk=0;
%%%%%%%%%%%%%%%%%%%%%%%%%%%%%%%%%
for k=1:1000;
Dz=[1 -a1 -a2];
R=roots(Dz);
rho(1,k)=abs(R(1)); theta(1,k)=angle(R(1));
rho(2,k)=abs(R(2)); theta(2,k)=angle(R(2));
alpha=0.98*alpha1;
yk1=yk;
yk=z(k);
xk2=xk1; xk1=xk;
xk=a1*xk1+a2*xk2;
sk3=sk2; sk2=sk1; sk1=sk;
sk=a1*sk1+a2*sk2+yk;
ek=yk-xk;
Jk=ek^2-v;
Obj(k)=Jk;
gk=[2*ek*sk1 2*ek*sk2 1]';
Hk_new=Hk/alpha;
delH_num=Hk_new*gk*gk'*Hk_new;
delH_den=1+gk'*Hk_new*gk;
delH=delH_num/delH_den;
Hk=Hk_new-delH;
old_delp_k=delp_k;
P(:,k)=pk;
pk=pk+delp_k;
a1=pk(1); a2=pk(2); v=pk(3);
Jk_hat=Jk-gk'*old_delp_k;
delp_k=Hk*gk*Jk_hat;
del_a1=delp_k(1);
del_a2=delp_k(2);
delx_k=sk1*del_a1+sk2*del_a2;
xk_corkt=xk+delx_k;
xk=xk_corkt;
alpha1=alpha1*0.9995+0.0005;
x(k)=xk;
y(k)=yk;
T(k)=k;
%%%
A(k)=a1; B(k)=a2; C(k)=v;
end
subplot(221);plot(T,0.5*x,T,y);grid;zoom;
subplot(222);plot(T,A,T,B,T,C);grid
subplot(212);polar (theta,rho,'o');grid;
```

A.5.4 MATLAB Program f51_3

```
%Digital Signal Processing:A Practitioner's Approach
%Dr.Kaluri Venkata Ranga Rao, kaluri@ieee.org
%Generates figure 5.14

clear; clf;
load f513.dat;
x1=f513(:,1);
y1=f513(:,2);
y2=f513(:,3);
t=1:length(y1);
s=20; e=90;
subplot(221)
plot(t,x1);grid;
title(' A ');xlabel(' Samples ');ylabel(' Input ');
subplot(222)
plot(t(s:e),y1(s:e),'.',t(s:e),y2(s:e));grid;
title(' B ');xlabel(' Samples ');ylabel(' Output ');
```

A.5.5 C Program f51_2

```c
# include <stdio.h>
# include <stdlib.h>
# include <math.h>
# include "f512matfun_f.c"
# include "f512dsp.c"

void main (void)
{
// test variables
int k;
int kmax = 2000;
float theta=20.0,f,yk,xk,pi;
float pk[3] = {0.1,0.1,0};
FILE *fp_write;

fp_write = fopen("f512.dat", "w");
f=theta/360.0;
pi = (atan(1.0))*4.0;
// randomise();
// printf(" start \n");
for (k=0; k < kmax; k++)
  {
```

```
yk=sin(2*pi*f*k) + 0.1*randn();
xk = m_rem(yk,pk);
// printf(" xk = %.3e yk = %.3e \n",xk,yk);
// getch();
fprintf(fp_write,"%.2f%.2f%.2f%.2f%.2f\n",pk[0],pk[1],pk[2],xk,yk);
  }
// printf(" finish \n");
fclose(fp_write);


}


void mat_p( float *,int,int);
void mat_mul( float *, float *, float *);
void mat_mul_col( float *, float *, float *);
void mat_mul_row( float *, float *, float *);
void mat_trps( float *, float *);
void mat_eql( float *, float *);
void ary_eql( float *, float *);
void mat_unt( float *);
void mat_scl( float *, float);
void ary_scl( float *, float);
void mat_mul_row_col( float *, float *, float *);
void mat_add( float *, float *, float *);
void ary_add( float *, float *, float *);
float mat_dot( float *, float *);
float randn(void);




void mat_mul( float *some_A, float *some_B, float *some_C)
{
int max_row_A = 3, max_col_A = 3;
int max_col_B = 3;
int col,row,k,index_A,index_B;
float val_A,val_B,sum;

for ( row = 0; row < max_row_A; row++)
{
for ( col = 0; col < max_col_B; col++)
{
sum = 0;
for ( k = 0; k < max_col_A; k++)
{
index_A = max_col_A*row + k;
```

```
index_B = max_col_B*k + col;

val_A = *(some_A + index_A);
val_B = *(some_B + index_B);

sum = sum + val_A*val_B;
}
*(some_C + max_col_A*row + col) = sum;
}
}
}



void mat_mul_row_col( float *some_A, float *some_B, float *some_C)
{
int max_col_C = 3, max_row_C = 3;
int col,row;
float val_A,val_B,sum;

for ( row = 0; row < max_row_C; row++)
{
for ( col = 0; col < max_col_C; col++)
{

val_A = *(some_A + row);
val_B = *(some_B + col);
sum = val_A*val_B;
*(some_C + max_col_C*row + col) = sum;
}
}
}

void mat_mul_col( float *temp_A, float *temp_B, float *temp_C)
{
int max_row_A = 3, max_col_A = 3;
int row,k,index_A,index_B;
float val_A,val_B,sum;

for ( row = 0; row < max_row_A; row++)
{

sum = 0;
for ( k = 0; k < max_col_A; k++)
```

```c
{
index_A = max_col_A*row + k ;
index_B = k ;

val_A = *(temp_A + index_A) ;
val_B = *(temp_B + index_B) ;

sum = sum + val_A*val_B ;
}
*(temp_C + row) = sum;
}
}


void mat_mul_row( float *temp_A, float *temp_B, float *temp_C)
{
int max_row_A = 3, max_col_A = 3;
int col,k,index_A,index_B;
float val_A,val_B,sum;

for ( col = 0; col < max_row_A; col++)
{

sum = 0;
for ( k = 0; k < max_col_A; k++)
{
index_A = max_col_A*k + col;
index_B = k ;
val_A = *(temp_A + index_A);
val_B = *(temp_B + index_B);

sum = sum + val_A*val_B ;
}
*(temp_C + col) = sum;

}
}


void mat_trps( float *some_A, float *some_B)
{
int max_row_A = 3, max_col_A = 3;
int max_col_B = 3;
int col,row,index_A,index_B;
```

```
float val_A;
for ( row = 0; row < max_row_A; row++)
{
for ( col = 0; col < max_col_B; col++)
{
index_A = max_col_A*row + col;
index_B = max_col_B*col + row;

val_A = *(some_A + index_A);
*(some_B + index_B) = val_A;
}
}
}




void mat_eql( float *some_A, float *some_B)
{
int max_row_A = 3, max_col_A = 3;
int max_col_B = 3;
int col,row,index_A,index_B;
float val_A;

for ( row = 0; row < max_row_A; row++)
{
for ( col = 0; col < max_col_B; col++)
{
index_A = max_col_A*row + col;
index_B = max_col_B*row + col;

val_A = *(some_A + index_A);
*(some_B + index_B) = val_A;
}
}
}




void ary_eql( float *some_A, float *some_B)
{
int max_row_A = 3;
int row;
float val_A;
```

```c
for (row = 0; row < max_row_A; row++)
{



val_A = *(some_A + row);
*(some_B + row) = val_A;
}
}



void ary_scl( float *some_A, float some_x)
{
int max_row_A = 3;
int row;
float val_A;
for ( row = 0; row < max_row_A; row++)
{

val_A = *(some_A + row);
*(some_A + row) = val_A*some_x;
}
}


void ary_add( float *some_A, float *some_B, float *some_C)
{
int max_row_A = 3;
int row;
float val_A,val_B;
for ( row = 0; row < max_row_A; row++)
{


val_A = *(some_A + row);
val_B = *(some_B + row);
*(some_C + row) = val_A + val_B;
}
}



void mat_add( float *some_A, float *some_B, float *some_C)
{
```

```
int max_row_A = 3, max_col_A = 3;
int max_col_B = 3;
int col,row,index_A,index_B;
float val_A,val_B;
for ( row = 0; row < max_row_A; row++)
{
for ( col = 0; col < max_col_B; col++)
{
index_A = max_col_A*row + col;
index_B = max_col_B*row + col;

val_A = *(some_A + index_A);
val_B = *(some_B + index_B);
*(some_C + index_A) = val_A + val_B;
}
}
}




void mat_scl( float *some_A, float b)
{
int max_row_A = 3, max_col_A = 3;
int col,row,index_A;
float val_A;

for ( row = 0; row < max_row_A; row++)
{
for ( col = 0; col < max_col_A; col++)
{
index_A = max_col_A*row + col;

val_A = *(some_A + index_A);
*(some_A + index_A) = val_A*b;
}
}
}




void mat_unt( float *some_A)
{
int max_row_A = 3, max_col_A = 3;
int col,row,index_A;
```

```
float val_A;

for ( row = 0; row < max_row_A; row++)
{
for ( col = 0; col < max_col_A; col++)
{
index_A = max_col_A*row + col;
val_A = 0;
if ( col == row ) val_A =1;
*(some_A + index_A) = val_A;
}
}
}



float mat_dot( float *temp_A, float *temp_B)
{
int max_row_A = 3;
int col,k,index_A,index_B;
float val_A,val_B,sum;
sum = 0.0;
for ( col = 0; col < max_row_A; col++)
{
index_A = col;
index_B = col;

val_A = *(temp_A + index_A);
val_B = *(temp_B + index_B);

sum = sum + val_A*val_B;
}
return(sum);
}


float randn(void)
{
float intmax,udf,sum;
int rnd,k;

intmax = 64000;
sum = 0.0;
for (k = 0; k < 12; k++)
```

```
{
rnd = rand();
udf = rnd;
udf = (udf/intmax)-0.5;
sum = sum + udf;
}
return(sum);
}



void mat_p( float *some_mat, int max_row, int max_col)
{
int col,row,index;
float val;
for (row =0; row < max_row; row++)
{
for (col =0; col < max_col; col++)
{
index = max_col*row + col;
val = *(some_mat + index);
printf (" %2.3e ",val);
}
printf("\n");
}
printf("\n");
}

float m_rem(float , float *);

float m_rem(float yk, float *pk)
{



static float xk=0, xk1=0.1, xk2;
static float sk=0, sk1=0, sk2=0, sk3, xk_corkt=0;

static float Hk[3] [3] = {{1e-9,0,0},{0, 1e-9,0},{0,0,1e-9}};
static float Hk_new[3][3],delH_num[3][3],delH[3][3],delH_den;

static float ek=0,Jk=0, Jk_hat=0;

static float gk[3] = {0,0,0};
```

```
static float delp_k[3]={0, 0, 0},old_delp_k[3]={0, 0, 0};
static float alpha,alpha1=0.95, delx_k=0;
static float temp1[3][3],temp2[3] [3];
static float scalar;
alpha=0.98*alpha1;

xk2=xk1;  xk1=xk;
xk=pk[0]*xk1+pk[1]*xk2;

sk3=sk2;  sk2=sk1;  sk1=sk;
sk=pk[0]*sk1+pk[1]*sk2+yk;

ek=yk-xk;

Jk=ek*ek-pk[2];

gk[0]=2*ek*sk1;
gk[1]=2*ek*sk2;
gk[2]= 1.0;

mat_eql(Hk,Hk_new);
scalar = 1.0/alpha;
mat_scl(Hk_new,scalar);
mat_mul_row_col(gk,gk,temp1);
mat_mul(Hk_new,temp1,temp2);
mat_mul(temp2,Hk_new,delH_num);


mat_mul_col(Hk_new,gk,temp1);
delH_den = 1 + mat_dot(gk,temp1);


mat_eql(delH_num,delH);
scalar = (-1.0/delH_den);
mat_scl(delH,scalar);

mat_add(Hk_new,delH,Hk);


ary_eql(delp_k,old_delp_k);
ary_add(pk,delp_k,pk);

Jk_hat=Jk - mat_dot(gk,old_delp_k)
mat_mul_col(Hk,gk,temp1);
```

```
ary_scl(temp1,Jk_hat);
ary_eql(temp1,delp_k);

delx_k=sk1*delp_k[0] + sk2*delp_k[1];

xk_corkt=xk+delx_k;

xk=xk_corkt;

alpha1=alpha1*0.9995 + 0.0005;

return(xk);
}
```

```
%Digital Signal Processing:A Practitioner's Approach
%Dr.Kaluri Venkata Ranga Rao, kaluri@ieee.org

close;clear
! f51_2main
load f512.dat;
  mrem=f512;
p1 = mrem(:,1);
p2 = mrem(:,2);
p3 = mrem(:,3);
p4 = mrem(:,4);
p5 = mrem(:,5);
k_c_code = length(p1);
%%%%%%%%%%%%%%%%%%%%%%%%%%%%%
for k=1:k_c_code;
Dz=[1 -p1(k) -p2(k)];
R=roots(Dz);
rho(1,k)=abs(R(1)); theta(1,k)=angle(R(1));
rho(2,k)=abs(R(2)); theta(2,k)=angle(R(2));
x(k)=p4(k); % Output Signal
y(k)=p5(k); % Input Signal
A(k)=p1(k);
B(k)=p2(k);
C(k)=p3(k);
T(k)=k;
%%%
end
  s=800; e=850;
```

```
subplot(221);plot(T(s:e),x(s:e),'.',T(s:e),x(s:e),T(s:e),
  y(s:e));grid;
xlabel(' Iterations ')
subplot(222);plot(T,A,T,B,T,C);grid
xlabel(' Iterations ');
ylabel(' Parameters ');
% subplot(212);polar(theta,rho,'o') ;grid;
```

A.6 Chapter 6 MATLAB Programs

In this chapter we gave a detailed treatment of a specific problem in direction finding, showing the need to understand systems engineering in order to implement DSP algorithms. It is very difficult to isolate DSP problems from other areas of engineering. Practical problems tend to cover many areas but centre on one area that demands the principal focus.

A.6.1 Program f6_9b

```
%Digital Signal Processing:A Practitioner's Approach
%Dr.Kaluri Venkata Ranga Rao, kaluri@ieee.org
%Generates figure 6.4


clear;close;
N=50;psi_deg=30;
k=1:N; f=0.1; theta=30; A_x=sin(theta*pi/180);A_y=cos(theta*pi/180);
psi_x=fix(rand(size(k))*psi_deg)*pi/180;
psi_y=fix(rand(size(k))*psi_deg)*pi/180;
x=A_x*cos(2*pi*f*k+psi_x);
y=A_y*cos(2*pi*f*k+psi_y);
subplot(221);
plot(x,y,'-');grid;title('A');xlabel('Analog');
%ylabel('-Y-');
u=ones(size(x));
A(:,1)=x';A(:,2)=u';
p=inv(A'*A)*A'*y';
X=linspace(min(x),max(x),100);
Y=p(1)*X+p(2);
subplot(222);
plot(x,y,'o',X,Y);grid;title('B');xlabel('Digital');
pause
print -depsc f6_9b
%ylabel('-Y-');
```

A.6.2 Program f6_3

```
%Digital Signal Processing:A Practitioner's Approach
%Dr.Kaluri Venkata Ranga Rao, kaluri@ieee.org
%Generates figures 6.1,6.2

clear;
rot=2;
for i=1:3
rot=rot-1;
theta=0:180;
theta=theta*pi/180;
r=(sin(theta)).^0.25;
phi=theta + rot*pi/4;
%polar (phi, r);grid;pause;
X(:,i)=phi'; Y(:,i)=r';
end
polar(X,Y);grid;pause;
%print -depsc f6_3
A=ones([271 3]);A=A*0;
A(1:181,1)=Y(1:181,1);
A(46:226,2)=Y(1:181,2);
A(91:271,3)=Y(1:181,3);
for i=1:271
  s=3*i-2;e=s+2;
  x(s:e)=A(i,:);
end
test=A(70,:); test=test+0.01*randn(size(test));
for i=1:271;
  ek=A(i,:)-test;
  Jk(i)=log(ek*ek');
end
subplot(211);plot(Jk);grid;
```

A.7 Library of Subroutines

This section gives subroutines needed by programs in the other sections.

A.7.1 Program Expand

```
%Digital Signal Processing:A Practitioner's Approach
%Dr.Kaluri Venkata Ranga Rao, kaluri@ieee.org
function y=expand(x,r)
% function y=expand(x,r)
```

```
% introduces r zeros between
% two succesive samples
% e:\matlab\toolbox\rf
   j=1:length(x);
k=j*r-r+1;
n=length(k);
y=zeros([1 n*r]);
y(k)=x;
clear k; clear j;
return;
```

A.7.2 Program freq

```
%Digital Signal Processing:A Practitioner's Approach
%Dr.Kaluri Venkata Ranga Rao , kaluri@ieee.org

  function[m,f]=freq(r);
 t=40*pi/180;
 b=[1 0 -1];
 a=[1 -2*r*cos(t) r*r];
 [h,w]=freqz(b,a,180);
 f=w/(2*pi);
 m=abs(h);
 p=angle(h);
 return
```

A.7.3 Program freqt

```
%Digital Signal Processing:A Practitioner's Approach
%Dr.Kaluri Venkata Ranga Rao, kaluri@ieee.org
  function[m,f]=freqt(th);
 r=0.9;
 t=th*pi/180;
 b=[1 0 -1];
 a=[1 -2*r*cos(t) r*r];
 [h,w]=freqz(b,a,180);
 f=w/(2*pi);
 m=abs(h);
 p=angle(h);
 plot(f,m)
 grid
 return
```

A.7.4 Program rbn_sig

```
%Digital Signal Processing:A Practitioner's Approach
%Dr.Kaluri Venkata Ranga Rao,kaluri@ieee.org
%This program needs rbn_sig.m, expand.m
function yy=rbn_sig(N,x)
% generates a random binary telegraph signal
% with +1 or 0
% where N = pulse width (samples)
%   and x is the binary sequence with +1 and -1
%   ( bi-polar is a must)
% x=sign(randn( [1 n]));
%  usage function yy=rbn_sig(N,x)
y=expand(x,N);
clear x;
k=find( y > 0 );
j=diff(k); i=find (j == N);
khat=k(i+1);
k=find( y < 0 );
j=diff(k); i=find (j == N);
khat_n=k(i+1);
y(khat)=y(khat)*0;
y(khat_n)=y(khat_n)*0;
b=1; a=[1 -1];
subplot(211)
plot(y);pause
yy=abs(filter(b,a,y));
subplot(212)
plot(yy)
clear y; clear k; clear khat_n;
clear khat; clear j; clear t;
clear a; clear b; clear i;
return
```

A.7.5 Program p2f1

```
%Digital Signal Processing:A Practitioner's Approach
%Dr.Kaluri Venkata Ranga Rao,kaluri@ieee.org

function [a,b,F]=p2f1(tow,f,A,n)
% function [a,b,F]=p2f (tow,f, A,n)
% converts a given pulse into
% fourier coefficients
% 'a' is a vector of cosine coefficeints
```

```
% 'b' is a vector of sine coefficents.
% 'F' is a vector giving the corresponding freq.
% 'tow' is the pulse width.
% 'f' is the frequency or the reciprocal of time-period.
% 'A' is the amplitude of the pulse.
% 'n' desired number of coefficients.
N=fix(n/2);
t=-N:1:N-1;
tc=pi*f*tow; F=t*f;
ta=A/pi;
a=ta*(sin(2*tc*t))./t;
b=2*ta*((sin(tc*t)).^2)./t;
   %plot(b);pause;
i=find(isnan(a));
v1=a(i+1); v2=a(i-1);
a(i)=(v1+v2)/2;
i=find(isnan(b));
v1=b(i+1); v2=b(i-1);
b(i)=(v1+v2)/2;
clear v1; clear v2; clear i;
return
```

A.7.6 Program pulse

```
%Digital Signal Processing:A Practitioner's Approach
%Dr.Kaluri Venkata Ranga Rao, kaluri@ieee.org
%This program needs pulse.m
function [y,delt]=pulse(tow,T,n);
% function [y,delt]=pulse(tow,T,n);
% generates a pulse of pulse width
% 'tow' sec and time period of 'T' sec
% of 'n' samples in the array 'y'
delt=T/n; n1=round(tow/delt);
t1=1:n1; y=1:n; y=0*y;
y(1:n1)=ones(size(t1));
return
```

A.7.7 Program forier

```
%Digital Signal Processing:A Practitioner's Approach
%Dr.Kaluri Venkata Ranga Rao , kaluri@ieee.org

function [f,a,b]=forier(x,delt);
% usage function [f,a,b]=forier(x,delt);
```

```
% This finds the fourier coefficents of the
% given periodic signal 'x'
% in dir d:\matlab\toolbox\rf\forier.m
Nx=length(x); frac=log(Nx)/log(2);
if ( (frac - fix(frac)) > 0 ) m=fix(frac)+1; Nx=2^m; end;
A=fftshift(fft(x,Nx));
n=round(length(A)/2);
a=real(A)/n; b=-imag(A)/n;
% A=fft(x); n=round(length(A)/2);
% a=(real(A(1:n)))/n; b=-(imag(A(1:n)))/n;
f=linspace(-0.5,0.5,length(A)); f=f/delt;
clear A;
return
```

A.7.8 Program spk

```
%Digital Signal Processing:A Practitioner's Approach
%Dr.Kaluri Venkata Ranga Rao, kaluri@ieee.org

function [sx,f]=spk(x,delt)

% usage function [sx,f]=spk(x,delt);
% This finds the spectrum of the given
% vector x
% returns 'sx' spectrum and 'f' the frequency
% in dir e:\matlab\toolbox\rf\spk.m
n=length(x);
frac=log(n)/log(2);
if ( (frac - fix(frac)) > 0 )
   m=fix(frac)+1; N=2^m;
else
    N=n;
   end;
% w=hamming(n);
% x1=w.*x';
sxtemp=fft(x,N);  f=linspace(0,0.5,N/2);
f=f/delt;
sx=(abs(sxtemp(1:N/2)))/(n/2);
return
```

A.7.9 Program pdf_y

```
%Digital Signal Processing:A Practitioner's Approach
%Dr.Kaluri Venkata Ranga Rao, kaluri@ieee.org
```

```
function [X,Y]=pdf_y(y,n)
% function [X,Y]=pdf_y(y,n)
% Finds the pdf of y specifed
% by n bins and returns the
% range and pdf in X and Y
% resectively
% e:\matlab\toolbox\rf
l=length(y);
[N,X]=hist(y,n);
Y=N/l;
clear l;
return;
```

A.8 Some Useful Programs

These programs demonstarte some of the key concepts for understanding the mathematics.

A.8.1 Program f2_grad

```
%Digital Signal Processing:A Practitioner's Approach
%Dr.Kaluri Venkata Ranga Rao, kaluri@ieee.org

clear;close;
x=-10:0.2:10;
t=1:length(x);
min_x=8;
ref=ones(size(x))*min_x;
y=2*(x-min_x).*(x-min_x)+6;
xk=1;
yk=2*(xk-min_x)*(xk-min_x)+6;
gk=4*(xk-min_x);
step = 1;
for i=1:length(x)+20
t(i)=i;ref(i)=min_x;
xl(i)=0;xl_prv(i)=0;xr(i)=0;xr_prv(i)=0;
xk1=xk;yk1=yk;gk1=gk;
xk=xk - step*yk/gk;
yk=2*(xk-min_x)*(xk-min_x)+6+0.2*randn;
gk=4*(xk-min_x)+0.1*randn;
xl(i)=xk;
yl(i)=yk;
change_in_g=sign(gk1)-sign(gk);
if change_in_g == 2
```

```
xk=xk1;
step=step/1.5;
end
if change_in_g == -2
xk=xk1;
step=step/1.5;
end
g(i)=change_in_g;
%if g(i) == 2
% xl(i)=xk;  xl_prv(i)=xk1;
%end
%if g(i) == -2
%  xr(i)=xk; xr_prv(i)=xk1;
%end
end
subplot(211);stem(t,g);grid;
subplot(212);plot(t,xl,t,ref);grid;
%pause;
%subplot(211);plot(t,xl_prv,'o',t,xl,'*',t,ref);grid
%subplot(212);plot(t,xr_prv,'o',t,xr,'*',t,ref);grid
```

A.8.2 Program f2_lemma

```
%Digital Signal Processing:A Practitioner's Approach
%Dr.Kaluri Venkata Ranga Rao, kaluri@ieee.org

clear;close
A=rand([3 3])-0.5; A=fix(A*20);
B=A+A';
c=rand([3 1])-0.5; c=fix(c*20);
delB=c*c';
Z=inv(B+delB);invB=inv(B);
Zhat=invB - invB*c*c'*invB/(1+c'*invB*c);
Z-Zhat
```

A.8.3 Program f5_pipe

```
%Digital Signal Processing:A Practitioner's Approach
%Dr.Kaluri Venkata Ranga Rao, kaluri@ieee.org

clear;clf ;k=1;
for i=30:1:50
theta=i*(pi/180);r=0.6;
```

```
p=2*cos(theta);
a3=r^3*(2*p-p^3); a4=r^4*(p*p-1);
b2=r*p;b3=r*r*(p*p-1);
a1=-r*p; a2=r*r;
a11=[1 0 0 a3 a4];
a11=1;
a22=1;
b22=[1 b2 b3];
a33=[1 a1 a2];
b33=1;
[h1 w]=freqz(b33,a33,180); w=w/(2*pi);
[h2 w]=freqz(b11,a11,180); w=w/(2*pi);
m1=abs (h1); m2=abs(h2);
rr=roots(a11); R=abs(rr);Theta=angle(rr);
A(:,k)=R;B(:,k)=Theta;
%polar(Theta,R,'o');grid;
%pause
k=k+1;
end
polar(B,A,'o');grid;
%plot(w,h1,'.-',w,h2);grid
```

A.8.4 Program f3_rate

```
%Digital Signal Processing:A Practitioner's Approach
%Dr.Kaluri Venkata Ranga Rao, kaluri@ieee.org

   close;clear;
N=3;
i=1:N;
a=[sum(i.*i) sum(i)
   sum(i)  N ];
b=inv(a);
  t=1:N; u = ones(size(t));
x= [4 5 6];
c=[t*x' u*x'] ';
mc=b*c
a1=1;
Nmax = 20;
b1 = 2*(1:Nmax);
b2 = (Nmax+1)*(ones(size(b1)));
    t1=1:200;
    noise=randn(size(t1))*6;
x1=sin(2*pi*0.01*t1);
```

```
x1=5*t1 + 1.5 + noise;
y1=filter(b1,a1,x1);
y2=filter(b2,a1,x1);
y=((y2-y1)*6)/((Nmax-1)*(Nmax+1)*Nmax);
subplot(211);
plot(t1,y,t1,x1);grid
subplot(212);
plot(t1,y);grid;
```

Index